Progress in Nonlinear Differential Equations and Their Applications

Volume 3

Michael Beals

Propagation and Interaction of Singularities in Nonlinear Hyperbolic Problems

1989

Birkhäuser
Boston · Basel · Berlin

Michael Beals
Department of Mathematics
Rutgers University
New Brunswick, NJ 08903, USA

Library of Congress Cataloging-in-Publication Data

Beals, Michael, 1954-
Propagation and interaction of singularities in nonlinear hyperbolic problems / Michael Beals.
p. cm.
Includes bibliographical references.
038734495
1. Wave equation--Numerical solutions. 2. Differential equations, Hyperbolic--Numerical solutions. 3. Nonlinear waves. 4. Singularities (Mathematics) I. Title.
QA927.B43 1989 530.1'24--dc20

89-17741

Printed on acid-free paper.

ISBN 0-8176-3449-5
ISBN 3-7643-3449-5

Camera-ready copy prepared by the authors using Macintosh Plus.
Printed and bound by Edwards Brothers, Inc., Ann Arbor, Michigan.
Printed in the USA.

9 8 7 6 5 4 3 2 1

To Phyllis

Preface

This book developed from a series of lectures I gave at the Symposium on Nonlinear Microlocal Analysis held at Nanjing University in October, 1988. Its purpose is to give an overview of the use of microlocal analysis and commutators in the study of solutions to nonlinear wave equations.

The weak singularities in the solutions to such equations behave up to a certain extent like those present in the linear case: they propagate along the null bicharacteristics of the operator. On the other hand, examples exhibiting singularities not present in the linear case can also be constructed. I have tried to present a crossection of both the regularity results and the singular examples, for problems on the interior of a domain and on domains with boundary. The main emphasis is on the case of more than one space dimension, since that case is treated in great detail in the paper of Rauch-Reed [59]. The results presented here have for the most part appeared elsewhere, and are the work of many authors, but a few new examples and proofs are given. I have attempted to indicate the essential ideas behind the arguments, so that only some of the results are proved in full detail. It is hoped that the central notions of the more technical proofs appearing in research papers will be illuminated by these simpler cases.

It is assumed that the reader is familiar with the basic notions of the classical theory of pseudodifferential operators, as presented in Folland [32], for example. The linear theory of the microlocal propagation of singularities and regularity is briefly presented here to motivate the nonlinear study.

I am grateful to Nanjing University, and especially Professor Qiu Qingjiu, for providing me the opportunity to lecture on these topics. In addition, I would like to thank Guy Métivier, Jeffrey Rauch, and Michael Reed for many years of enlightening discussions, and Mark Williams for his detailed comments on the manuscript. Research support was provided by the Alfred P. Sloan Foundation and the Rutgers University Research Council.

Contents

Introduction

We consider the regularity of a class of solutions to nonlinear strictly hyperbolic problems. The results apply to local solutions of semilinear equations of the form

$$(0.1) \qquad P_m(x,D_x)u = f(x,u,\ldots,D_x^{m-1}u),$$

where f is assumed to be a C^∞ function of its arguments and u is sufficiently smooth to make sense of the right hand side. For example, Sobolev regularity of the form $u \in H^{s+m-1}(\mathbb{R}^n)$ for $s > (n+1)/2$ is sufficient to guarantee the local existence of such solutions. Under more stringent assumptions on the regularity of u, many of the smoothness results to be considered have also been established for solutions of quasilinear equations

$$(0.2) \qquad \sum_{|\alpha|=m} a_\alpha(x,u,\ldots,D_x^{m-1}u)D^\alpha u = f(x,u,\ldots,D_x^{m-1}u),$$

or fully nonlinear equations of the form

$$(0.3) \qquad f(x,u,\ldots,D^\alpha u)_{|\alpha|\le m} = 0.$$

Pseudodifferential operators have been an important tool in the analysis of solutions to linear problems. They allow the decomposition of the singular part of solutions into pieces which can be analyzed separately and then reassembled. When a nonlinear function acts on a solution, the interaction of the microlocal pieces complicates the procedure, but in many cases the analysis can be successfully carried out. Beginning with the work of Rauch [57] and Bony [15], the tools of microlocal analysis have been applied to the study of nonlinear problems such as (0.1), (0.2) and (0.3) in an attempt to describe the propagation of singularities and of regularity for solutions. If the location and type of the singularities of a solution are known on an initial surface or in the past, the question is to determine the location and type of the singularities in the future. Certain results analogous to those of the linear theory continue to hold, but purely nonlinear phenomena are also known to occur. In this monograph we present a number of the developments in this nonlinear theory of microlocal singularities.

Several of the basic properties of solutions to linear wave equations are recalled in Chapter I. In particular, energy estimates in Sobolev spaces and characterizations of wave front sets are given. Microlocal Sobolev regularity of certain types is shown to be preserved under the action of nonlinear functions. This property allows the extension to solutions of (0.1), (0.2), and (0.3) of the linear result on the propagation of smoothness, up to a certain order of regularity. In Chapter II, examples of solutions to (0.1) are given which exhibit singularities not present in the solution to the corresponding linear problem $p_m(x,D_x)v = 0$ with the same Cauchy data as u. It is shown that nonlinear singularities in u can arise at the crossing points of singularities in v, and that a single singularity in v corresponding to a line in the wave front set can give rise to additional singularities in u. As a consequence, it is shown that a solution to (0.1) can have singularities on the largest set allowed by finite propagation speed, unlike the linear case.

More restricted types of regularity, which curtail the generation of many nonlinear singularities, are considered in Chapter III. Solutions of (0.1) or (0.2) which are conormal with respect to one or two smooth characteristic hypersurfaces for the equation in the past are shown to remain conormal with respect those hypersurfaces in the future if the order of the equation is two, and with respect to an appropriate family of characteristic hypersurfaces in the higher order case. Examples of conormal regularity with respect to singular hypersurfaces for (0.1) are also given. In Chapter IV it is shown that the interaction of conormal singularities on three characteristic hypersurfaces can generate an additional hypersurface of singularities for a solution to (0.1). A modification of the usual commutator argument is given to prove that the solution is conormal with respect to the family which includes the new hypersurface.

Many of the general regularity results established in Chapter I are extended in Chapter V to problems on domains with non-characteristic boundary. The appropriate spaces are now the Hörmander spaces, and tangential microlocalization is employed. Sobolev regularity up to a certain order is shown to propagate, even in the presence of singularities along grazing rays. Conormal solutions on domains with boundary are considered in Chapter VI. A solution conormal in the past with respect to a single characteristic hypersurface which intersects the boundary away from grazing directions is shown to remain conormal with respect to the reflected family of characteristic hypersurfaces.

Chapter I. Nonlinear Microlocal Analysis

The prototype of the equations described in the introduction is the simple semilinear wave equation

$$(1.1) \qquad \Box u = \Big\{\partial_t^2 - \sum_{j=1}^{n} \partial_{x_j}^2\Big\} u = f(t,x,u),$$

with f an arbitrary smooth function. We consider first the linear case

$$(1.2) \qquad \Box u = 0$$

and several elementary examples of solutions.

(a) If $H(r)$ is the Heaviside function in one variable,

$$H(r) = 1,\ r \geq 0, \quad H(r) = 0,\ r < 0,$$

then $u(t,x) = H(t - x_j)$ is perhaps the simplest solution of (1.2) which has a singularity. If the number of space dimensions n is larger than one, then the singular support of this solution and its time derivative at $t = 0$ is the codimension one set $\{x_1 = 0\}$; at any later time the solution also has singular support on a set of codimension one. See Figure 1.1a).

(b) Let $u(t,x) = \int e^{i(x\cdot\xi - t|\xi|)} d\xi$ in the sense of distributions, that is, for $\varphi \in C^\infty_{com}(\mathbf{R}^{n+1})$, $\langle u, \varphi \rangle = \iiint e^{i(x\cdot\xi - t|\xi|)} \varphi(t,x)\, dx dt d\xi$. The measure $d\xi$ in the dual variables will always be normalized to be $(2\pi)^{-n}$ times the usual Lebesgue measure. Then, as is easily verified, $u|_{t=0} = \delta_0$, the Dirac distribution at the origin, and by arguments similar to those which follow, $\partial_t u|_{t=0}$ has singular support at the origin. Moreover, the singular support of u is $\{(t,x): |t| = |x|\}$. This property may be verified by using a standard

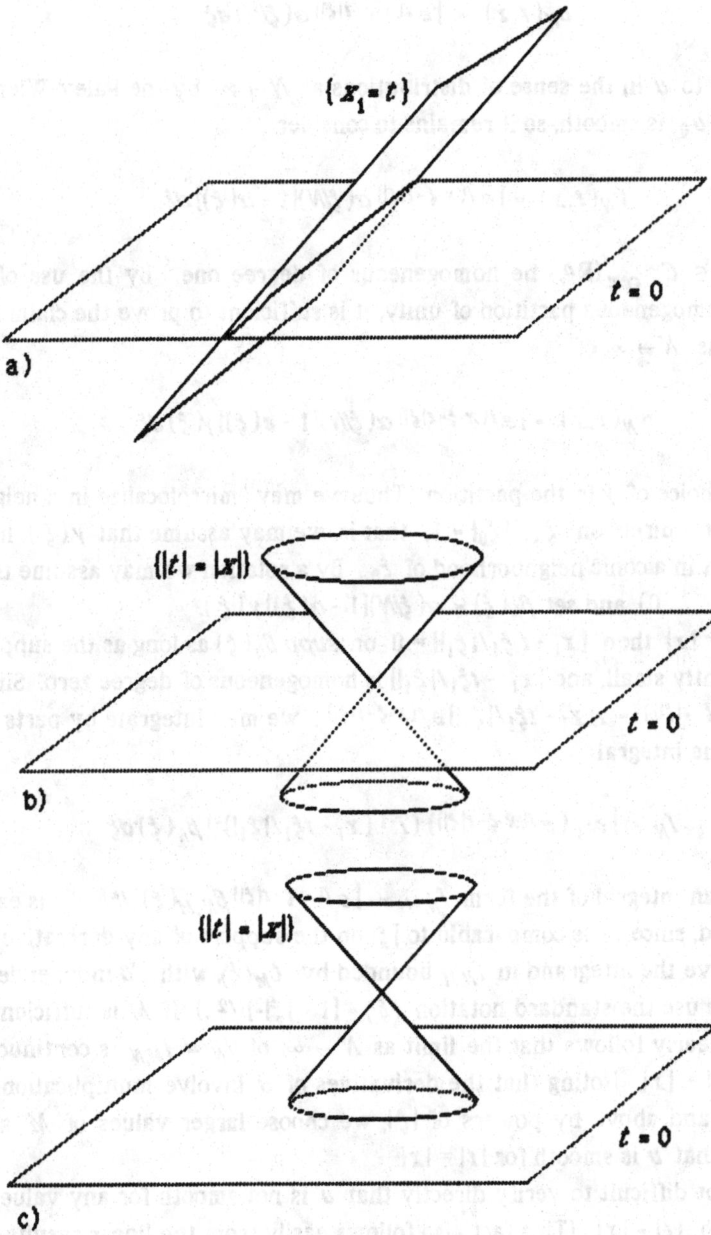

Figure 1.1

integration by parts argument which appears frequently in the analysis of singularities, so we sketch it here.

Let $\alpha \in C^{\infty}{}_{com}(\mathbf{R}^n)$, $\alpha(\xi) = 1, |\xi| \le 1$, $\alpha(\xi) = 0, |\xi| \ge 2$. Then

$$u_N(t,x) = \int e^{i(x\cdot\xi - t|\xi|)} a(\xi/N)\, d\xi$$

converges to u in the sense of distributions as $N \to \infty$. By the Paley-Wiener Theorem, u_0 is smooth, so it remains to consider

$$v_N(t,x) = \int e^{i(x\cdot\xi - t|\xi|)} a(\xi/N)[1 - a(\xi)]\, d\xi.$$

Let $\gamma(\xi) \in C^\infty_{com}(\mathbf{R}^n)$ be homogeneous of degree one. By the use of a smooth homogeneous partition of unity, it is sufficient to prove the claim for the limit as $N \to \infty$ of

$$w_N(t,x) = \int e^{i(x\cdot\xi - t|\xi|)} a(\xi/N)[1 - a(\xi)]\gamma(\xi)\, d\xi.$$

for such choice of γ in the partition. Thus we may "microlocalize in a neighborhood of a direction ξ_0, $|\xi_0| = 1$," that is, we may assume that $\gamma(\xi)$ has its support in a conic neighborhood of ξ_0. By a rotation we may assume that $\xi_0 = (1,0,\ldots,0)$, and set $\beta_N(\xi) = a(\xi/N)[1 - a(\xi)]\gamma(\xi)$.

If $|t| \neq |x|$ then $[x_1 - t\,\xi_1/|\xi_1|] \neq 0$ on $supp\, \beta_N(\xi)$ as long as the support is sufficiently small, and $[x_1 - t\xi_1/|\xi_1|]$ is homogeneous of degree zero. Since $\partial_{\xi_1}(e^{i(x\cdot\xi - t|\xi|)}) = i[x_1 - t\xi_1/|\xi_1|]e^{i(x\cdot\xi - t|\xi|)}$, we may integrate by parts M times in the integral

$$I_N = \int \partial_{\xi_1}(e^{i(x\cdot\xi - t|\xi|)})\,(i^{-1}(x_1 - t\xi_1/|\xi_1|)^{-1}\beta_N(\xi)\, d\xi$$

to obtain an integral of the form $I_{N,M} = \int e^{i(x\cdot\xi - t|\xi|)}\beta_{N,M}(\xi)\, d\xi$. As is easily verified, since N is comparable to $|\xi|$ on the support of any derivative of β_N, we have the integrand in $I_{N,M}$ bounded by $C_M\langle\xi\rangle$, with C_M independent of N. (We use the standard notation $\langle\xi\rangle = [1 + |\xi|^2]^{1/2}$.) If M is sufficiently large, it clearly follows that the limit as $N \to \infty$ of $I_N = I_{M,N}$ is continuous where $|t| \neq |x|$. Noting that the derivatives of u involve multiplication of the integrand above by powers of $|\xi|$, we choose larger values of M and conclude that u is smooth for $|t| \neq |x|$.

It is not difficult to verify directly that u is not smooth for any value of (t,x) with $|t| = |x|$. (This fact also follows easily from the linear results on propagation of singularities described subsequently, as well as the rotational and dilation symmetry of u.) Thus, even though the singular support of this solution and its time derivative at $t = 0$ is the codimension n set $\{0\}$, at any later time the solution has singular support on a set of codimension one. See

Figure 1.1b).

(c) Let $u(t,x) = \int e^{i(x\cdot\xi - (t-1)|\xi|)} d\xi$ in the sense of distributions. This solution is a time translate of the preceding one, as shown in Figure 1.1c). At time $t = 0$ its singular support has codimension one, while at time $t = 1$ its singular support has codimension n.

The question to be considered in general is the following: given initial data with prescribed set of singularities for a solution to a strictly hyperbolic problem, or given the location of the singularities of the solution in the past, what is the location of the singularities at a later time? The examples above show that even in the linear case, precise information may not follow merely from knowledge of the location of the initial singularities. By superposition of translates of solutions of the type given in b), one may use a "condensation of singularities" argument to obtain solutions u which at time zero are singular only on $\{x_1 = 0\}$ but which satisfy $\mathit{sing\,supp}(u) = \{(t,x): |x_1| \le |t|\}$. See Figure 1.2. Such functions u are in striking contrast to the solution given in a), which had initial singularities on the same set, or in c), where the size of the singular set is at one later time significantly less than its initial size. Moreover, one can find appropriate data (to be described in Chapter II) singular only at the origin, as in the solution given in b), but for which the solution to (1.2) is singular only on the line $\{(t,x): t = x_1, x' = 0\}$. General statements about the propagation of singularities in the linear case can be made, for instance by use of the property of finite propagation speed. However, refined statements require a more precise notion of the singular support of the function. The Fourier spectrum of a solution to the wave equation is intimately connected with its singularities. The wave front set, defined in the C^∞ case by Hörmander (see, for example, [36]), is the appropriate setting for analysis in the linear case.

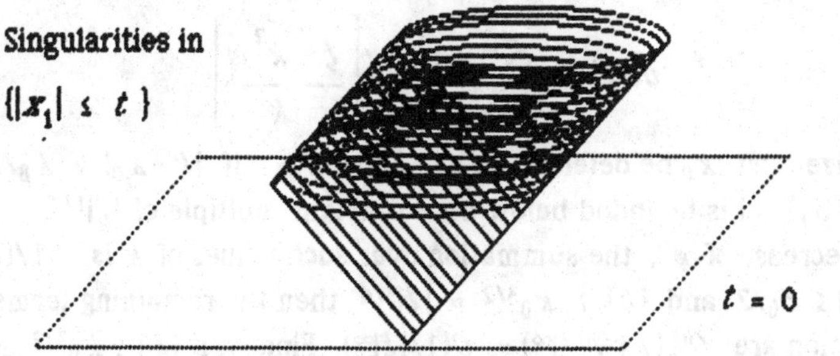

Figure 1.2

Definition 1.1. For a distribution u on $\mathbf{R}^n$, the wave front set is the subset of the cotangent space $T^*(\mathbf{R}^n)\backslash 0$ determined as follows: $(x_0, \xi_0) \notin WF(u)$ if there exist $\varphi \in C^\infty_{com}(\mathbf{R}^n)$ with $\varphi(x_0) \neq 0$ and a conic neighborhood $\Gamma \subset \mathbf{R}^n$ of ξ_0 such that $\chi_\Gamma(\xi)(\varphi u)^\wedge(\xi)$ is rapidly decreasing as $|\xi| \to \infty$. Here χ_Γ is the characteristic function of Γ.

As is easily verified, the Paley-Wiener Theorem implies that the wave front set of a distribution is a refinement of its singular support: $x_0 \in$ *sing supp*(u) if and only if there exists $\xi_0 \neq 0$ such that $(x_0, \xi_0) \in WF(u)$. This property simplifies the computation of the wave front set in examples. For instance, if $H(x)$ is the Heaviside function as defined in a), then the x projection of the wave front set is $\Pi_x WF(H) = \{x_1 = 0\}$. For any functions $\alpha \in C^\infty_{com}(\mathbf{R}^n)$, $\beta \in C^\infty_{com}(\mathbf{R}^{n-1})$, $(\alpha\beta H)^\wedge(x_1, x') = (\alpha H)^\wedge(\xi_1)\beta^\wedge(\xi')$. This function is rapidly decreasing for any $\xi = (\xi_1, \xi')$ with $\xi' \neq 0$, since $\langle \xi \rangle \approx \langle \xi' \rangle$ for such ξ and $\beta^\wedge$ is rapidly decreasing. Therefore $WF(H) = \{(0, x', \xi_1, 0)\colon \xi_1 \neq 0\}$. Similarly, if δ_0 is the Dirac distribution at the origin and if $\alpha(0) = 1$, then we have $(\alpha\delta_0)^\wedge(\xi) = 1$, and thus $WF(\delta_0) = \{(0, \xi)\colon \xi \neq 0\}$.

In [36], Hörmander constructs the fundamental example of a distribution u with wave front set consisting of a single ray in the cotangent space. It follows by superposition of such functions and a condensation of singularities argument (that is, use of the Baire Category Theorem) that given any closed conic subset of $T^*(\mathbf{R}^n)\backslash 0$, there is a function having that subset as its wave front set. Consider the case of a single ray, taken to be $\{(0, r\omega_1)\colon r > 0\}$, with $\omega_1 = (1, 0, \ldots, 0)$. Let $\varphi \in C^\infty_{com}(\mathbf{R}^n)$, $\varphi^\wedge(\xi) \geq 0$, $\varphi^\wedge(0) > 0$, and set

$$u(x) = \sum_{k=1}^{\infty} \frac{1}{k^2} e^{ik^2(x \cdot \omega_1)} \varphi(kx).$$

Notice that the sum is finite for $x \neq 0$, so u is smooth there, and the sum is absolutely convergent, so u is continuous. The Fourier transform is

$$\hat{u}(\xi) = \sum_{k=1}^{\infty} \frac{1}{k^{2+n}} \hat{\varphi}\left[\frac{\xi - k^2\omega_1}{k}\right].$$

For ξ fixed, let k_0 be determined by $|\xi| \approx k_0^2$. If $|k - k_0| \geq k_0/2$, then $|(\xi - k^2\omega_1)/k|$ is bounded below by a constant multiple of $|\xi|^{1/2}$, so by the rapid decrease of $\varphi^\wedge$, the summation over such values of k is $O(1/|\xi|^\infty)$. If $|k - k_0| \leq k_0/2$ and $|\xi'| \geq k_0^{3/2} \approx |\xi|^{3/4}$, then the remaining terms in the summation are $O([1/k_0^{3/2}]^\infty) = O(1/|\xi|^\infty)$. Finally, if $|\xi'| \leq k_0^{3/2}$ and ξ_1 is negative, then a similar estimate holds. Thus the only possible direction in

$\Pi_\xi WF(u)$ is the ray through ω_1. Since the singular support and hence the wave front of u is clearly nonempty, we have $WF(u) = \{(0, \tau\omega_1): \tau > 0\}$. Moreover, there are estimates on $u^\wedge$ from below on neighborhoods of the ray through ω_1 which are useful in applications. If, more generally, we consider the function

$$(1.3) \qquad u(x) = \sum_{k=1}^{\infty} \frac{1}{k^{1+\sigma}} e^{ik^{1+\rho}(x\cdot\omega_1)} \varphi(kx), \text{ for } \sigma > 0, \rho > 0,$$

the same arguments as above yield $WF(u) = \{(0, \rho\omega_1): \rho > 0\}$. By increasing the value of σ, the smoothness of u at the origin can be increased. The Fourier transform is given by

$$(1.4) \qquad u^\wedge(\xi) = \sum_{k=1}^{\infty} \frac{1}{k^{1+\sigma+n}} \varphi^\wedge\left[\frac{\xi - k^{1+\rho}\omega_1}{k}\right]$$

On a neighborhood of the form $\{\xi: \left|\xi/|\xi| - \omega_1\right| \le C|\xi|^{-O(\rho)}\}$, it is not difficult to obtain lower bounds on $u^\wedge$ of the type $u^\wedge(\xi) \ge C|\xi|^{1+\sigma+n+O(\rho)}$ as $\rho \to 0$. By choosing ρ near zero one obtains the largest such bounds from below compatible with the condition that the wave front set of u contains only one ray. (If such an estimate held on a neighborhood of the form $\{\xi: \left|\xi/|\xi| - \omega_1\right| \le C\}$, then all of the directions ξ in that neighborhood would be contained in $\Pi_\xi WF(u)$.)

The wave front set is well behaved under the standard operations of linear analysis. For example, it is easily verified that

$$WF(u_1 + u_2) \subset WF(u_1) \cup WF(u_2),$$

$WF(\alpha u) \subset WF(u)$ for any smooth function α, and

$WF(p(x,D)u) \subset WF(u)$ for any smooth differential operator $p(x,D)$.

The application of a nonlinear function to u may complicate matters, though, since the Fourier transform of the product of two functions is the convolution of the Fourier transforms. In general, the wave front set will only behave in a satisfactory fashion when $\Pi_\xi WF(u) \subset K_1$, a closed convex proper subcone of $\mathbf{R}^n \setminus 0$. For example, if the distribution u has this property, then u^2 is well defined, and $\Pi_\xi WF(u^2) \subset \Pi_\xi WF(u)$. More generally, if $\Pi_\xi WF(u_i) \subset K_i$ for $i = 1, 2$, and $\xi \in K_1$ implies that $-\xi \notin K_2$, then $u_1 u_2$ is well defined and $\Pi_\xi WF(u_1 u_2) \subset \{K_1 + K_2\}^{closure}$. Indeed, without loss of generality it may be assumed that $supp(u_i)$ is compact, so that

$$(u_1 u_2)^\wedge(\xi) = \int u_1^\wedge(\xi - \eta) u_2^\wedge(\eta)\, d\eta .$$

The integrand may be decomposed by writing

$$\hat{u_j}(\xi) = \chi_{K_j}(\xi)\hat{u_j}(\xi) + \chi_{K_j^{comp}}(\xi)\hat{u_j}(\xi).$$

The integrals involving $K_j{}^{comp}$ are easily estimated, because of the rapid decrease of u_j on these cones. The integral involving $\chi_{K_1}\chi_{K_2}$ is easily seen to be supported where $\xi \in (K_1 + K_2)^{closure}$, and for each fixed ξ, that integral is taken over a compact set. (If $|\xi| \le \varepsilon|\eta|$, then $(\xi - \eta) \approx -\eta$, while $(\xi - \eta) \in K_1$ and $\eta \in K_2$, contradicting the assumptions on K_1 and K_2, and therefore $|\eta| \le |\xi|/\varepsilon$ on the support of this integrand.) For details of this argument, see Hörmander [35].

Without some conditions on the wavefronts, it is not possible in general to make sense of the product of two distributions. For example, if $u = \delta_0$, then u^2 cannot be reasonably defined, since if $\varphi \in C^\infty{}_{com}(\mathbf{R}^n)$ has integral equal to one, and $\varphi_k(x) = k^n\varphi(kx)$, then $\varphi_k(x) \to \delta_0$ as $k \to \infty$, but $\varphi_k{}^2(x)$ does not converge in the sense of distributions. However, if u is sufficiently smooth, nonlinear functions of u are well defined. We will measure regularity in the Sobolev spaces, since this type of smoothness is preserved under the action of operators associated with linear strictly hyperbolic equations.

Definition 1.2. $H^s(\mathbf{R}^n) = \{u: \langle\xi\rangle^s u^\wedge(\xi) \in L^2(\mathbf{R}^n)\}$. For $u \in H^s(\mathbf{R}^n)$, set $\|u\|_s = \|\langle\xi\rangle^s u^\wedge(\xi)\|_{L^2(\mathbf{R}^n)}$. Similarly, $H^s{}_{loc}(\mathbf{R}^n) = \{u: \varphi u \in H^s(\mathbf{R}^n)$ for all $\varphi \in C^\infty{}_{com}(\mathbf{R}^n)\}$.

If the regularity index s is large enough, nonlinear analysis is appropriate on the Sobolev space $H^s(\mathbf{R}^n)$. For example, if $s > n/2$, then the Sobolev Embedding Theorem implies that $H^s(\mathbf{R}^n) \subset L^\infty(\mathbf{R}^n)$. (See, for example, Stein [69].) Moreover, $H^s(\mathbf{R}^n)$ is an algebra for $s > n/2$.

Lemma 1.3 (SCHAUDER). *If* $u, v \in H^s(\mathbf{R}^n)$ *and* $s > n/2$, *then* $uv \in H^s(\mathbf{R}^n)$, *and* $\|uv\|_{H^s} \le C\|u\|_{H^s}\|v\|_{H^s}$.

Proof. We may write

$$\langle\xi\rangle^s(uv)^\wedge(\xi) = \int\langle\xi\rangle^s u^\wedge(\xi-\eta)v^\wedge(\eta)\,d\eta$$
$$= \int\langle\xi\rangle^s\langle\xi-\eta\rangle^{-s}\langle\eta\rangle^{-s} f^\wedge(\xi-\eta)g^\wedge(\eta)\,d\eta ,$$

with $f, g \in L^2(\mathbb{R}^n)$. If $|\xi - \eta| \le |\xi|/2$, then $|\eta| \ge |\xi|/2$, and therefore $\langle\xi\rangle^s\langle\xi-\eta\rangle^{-s}\langle\eta\rangle^{-s} \le C\langle\xi-\eta\rangle^{-s}$. On the other hand, if $|\xi-\eta| \ge |\xi|/2$, then $\langle\xi\rangle^s\langle\xi-\eta\rangle^{-s}\langle\eta\rangle^{-s} \le C\langle\eta\rangle^{-s}$. The result now follows from Lemma 1.4 below. Q.E.D.

The following result, which will be useful in a variety of contexts, may be found in Rauch-Reed [59].

Convolution Lemma 1.4. *Suppose that $K(\xi,\eta)$ may be decomposed into finitely many pieces $K_j(\xi,\eta)$, each of which satisfies either*

$$\sup_\xi \int |K_j|^2\, d\eta \le C_0 < \infty \quad \text{or} \quad \sup_\eta \int |K_j|^2\, d\xi \le C_0 < \infty .$$

Then for $f, g \in L^2(\mathbb{R}^n)$ and $h(\xi) = \int K(\xi,\eta) f(\xi-\eta) g(\eta)\, d\eta$, it follows that $h \in L^2(\mathbb{R}^n)$ and $\|h\|_{L^2(\mathbb{R}^n)} \le cC_0 \|f\|_{L^2(\mathbb{R}^n)} \|g\|_{L^2(\mathbb{R}^n)}$.

Proof. Let $k \in L^2(\mathbb{R}^n)$, $\|k\|_{L^2(\mathbb{R}^n)} \le 1$, and then apply the Cauchy-Schwarz inequality to obtain

$$\begin{aligned}\|h\|_{L^2(\mathbb{R}^n)} = \sup\left|\int h(\xi)k(\xi)\,d\xi\right| &\le \min\{ [\textstyle\iint |K(\xi,\eta)|^2 |g(\eta)|^2\, d\xi d\eta]^{1/2} [\iint |f(\xi-\eta)|^2 |k(\xi)|^2\, d\eta d\xi]^{1/2}, \\ &\qquad [\textstyle\iint |K(\xi,\eta)|^2 |k(\xi)|^2\, d\eta d\xi]^{1/2} [\iint |f(\xi-\eta)|^2 |g(\eta)|^2\, d\xi d\eta]^{1/2} \}.\end{aligned}$$

The result follows by decomposing $K(\xi,\eta)$. Q.E.D.

Schauder's Lemma may be generalized in a number of ways. For example, the same type of proof yields the following estimate on products of functions in lower order Sobolev spaces. See Beals [7].

$$\text{(1.5)} \qquad \begin{aligned}&\text{If } u \in H^{s_1}(\mathbb{R}^n),\ v \in H^{s_2}(\mathbb{R}^n), \text{ with } s_1, s_2 \ge 0, \text{ then} \\ &uv \in H^{\min(s_1, s_2, s_1 + s_2 - n/2 - \delta)} \quad \text{for any } \delta > 0 .\end{aligned}$$

Moreover, the multiplication of functions of differing Sobolev regularity allows a simple argument extending the invariance under polynomials of the spaces $H^s_{loc}(\mathbb{R}^n)$ for $s > n/2$, which is a corollary of induction and Lemma 1.3, to invariance under the action of smooth functions.

Lemma 1.5. *If $u \in H^s_{loc}(\mathbb{R}^n)$ for $s > n/2$, and if $f(x,v)$ is a smooth function of its arguments, then $f(x,u) \in H^s_{loc}(\mathbb{R}^n)$.*

Proof. Without loss of generality, it may be assumed that u and f have compact support, with $f(0) = 0$. Then by the Sobolev Embedding Theorem, $f(x,u) \in L^\infty(\mathbf{R}^n)$, and the compact support implies that $f(x,u) \in H^0(\mathbf{R}^n)$. Suppose we have established inductively that for any smooth function g and for some r with $0 < r < s$, $g(x,u) \in H^r(\mathbf{R}^n)$. Let D_x stand for any smooth vector field; then the chain rule (as long as it can be shown to apply) yields that $D_x[f(x,u)] = [D_x f](x,u) + [D_u f](x,u) D_x u$. By the inductive hypothesis, $[D_x f](x,u) \in H^r(\mathbf{R}^n)$, while the inductive hypothesis and (1.5) imply that $[D_u f](x,u) D_x u \in H^{min(r, s-1, r+s-1-n/2-\delta)}(\mathbf{R}^n)$. (If $n = 1$, a slight modification of (1.5) is necessary.) It follows easily by the norm estimates corresponding to (1.5) that the use of the chain rule is justified. Then, for $\varepsilon = min(1, [s-n/2]/2)$, $D_x[f(x,u)] \in H^{min(r-1+\varepsilon, s-1)}(\mathbf{R}^n)$, and therefore $f(x,u) \in H^{min(r+\varepsilon, s)}(\mathbf{R}^n)$. The proof is completed by induction on r. Q.E.D.

The fundamental estimate satisfied by solutions of linear strictly hyperbolic equations is the so-called energy inequality. It describes the Sobolev regularity of a solution to $p(x,D)u = f(x)$ in terms of the regularity of f and the regularity of the restriction of u to a space-like initial surface or to a region in the past which determines the values of u in the future. One version is the following; finite propagation speed would allow the replacement of global hypotheses with suitable local ones. For a proof, see, for example, Chazarain-Piriou [21].

Linear Energy Inequality 1.6. *Let $p(x,D)$ be a partial differential operator of order m on $\mathbf{R}^{n+1}$, strictly hyperbolic with respect to the planes $\{x_{n+1} = c\}$, and let u satisfy $p(x,D)u = f(x)$. If $f \in H^{s-m+1}_{loc}(\mathbf{R}^{n+1})$ and $u \in H^s_{loc}(\{x: |x_{n+1}| < e\})$ for some $\varepsilon > 0$, then $u \in H^s_{loc}(\mathbf{R}^{n+1})$.*

This result may be made more precise by considering the "microlocal" Sobolev spaces, which measure the L^2 size of the Fourier transform on cones as in the definition of wave front set. It is this description of the regularity which allows the most precise determination of how regularity and singularities for solutions to linear strictly hyperbolic equations are propagated.

Definition 1.7. $u \in H^s_{ml}(x_0, \xi_0)$ if there exist $\varphi \in C^\infty_{com}(\mathbf{R}^n)$, $\varphi(x_0) \neq 0$, and a conic neighborhood $\Gamma \subset \mathbf{R}^n$ of ξ_0 such that $\langle \xi \rangle^s \chi_\Gamma(\xi)(\varphi u)\hat{}(\xi) \in L^2(\mathbf{R}^n)$.

There is an equivalent definition in terms of classical pseudodifferential operators. Recall, we say that $a(x,\xi)$ is a classical symbol of order m, and write $a(x,\xi) \in S^m{}_{1,0}$, if for all compact sets K and for all multiindices there are constants $C_{K,\alpha,\beta}$ for which $|D_x{}^\alpha D_\xi{}^\beta a(x,\xi)| \le C_{K,\alpha,\beta}\langle\xi\rangle^{m-|\beta|}$ for all $x \in K$. The corresponding operator, defined for instance on $C^\infty{}_{com}(\mathbf{R}^n)$, is given by

$$a(x,D)u(x) = \int e^{ix\cdot\xi}\, a(x,\xi)\, u\hat{\ }(\xi)\, d\xi.$$

If $a(x,\xi) \in S^m{}_{1,0}$ and $u \in H^s{}_{loc}(\mathbf{R}^n)$, then $a(x,D)u(x) \in H^{s-m}{}_{loc}(\mathbf{R}^n)$. The symbolic calculus for such operators is well known. For example, a symbol is said to be microlocally elliptic at (x_0,ξ_0) if there is a conic neighborhood of (x_0,ξ_0) and a constant c such that on that neighborhood, $|a(x,\xi)| \ge c\langle\xi\rangle^m$. Such microlocally elliptic operators may be inverted near their elliptic points. The notions of wavefront set and microlocal regularity may be reexpressed in terms of the actions of pseudodifferential operators (see Hörmander [37]). As is easily verified from the definitions and the symbolic calculus, $(x_0,\xi_0) \notin WF(u)$ if and only if there is a classical pseudodifferential operator with symbol $a(x,\xi)$ microlocally elliptic at (x_0,ξ_0) such that $a(x,D)u(x)$ is rapidly decreasing as $|\xi| \to \infty$. Moreover, $u \in H^s{}_{ml}(x_0,\xi_0)$ if and only if there is a classical pseudodifferential operator of order zero with symbol $a(x,\xi)$ microlocally elliptic at (x_0,ξ_0) such that $a(x,D)u(x) \in H^s{}_{loc}(\mathbf{R}^n)$.

The microlocal reformulation of the linear energy inequality is the statement that microlocal regularity is propagated along the null bicharacteristics for the strictly hyperbolic operator. Recall that for points with $p(x_0,\xi_0) = 0$, the null bicharacteristic through (x_0,ξ_0) is the curve defined by the Hamiltonian system

$$dx/ds = \nabla_\xi\, p(x,\xi), \quad d\xi/ds = -\nabla_x\, p(x,\xi), \quad x(0) = x_0, \ \xi(0) = \xi_0.$$

There is a unique solution to these equations when p is strictly hyperbolic, and $p(x,\xi) \equiv 0$ along such a curve.

As an example we consider the ordinary wave operator $\Box$, with symbol $\tau^2 - |\xi|^2$. The null bicharacteristic through the point $(0,x_0,\tau_0,\xi_0)$ with $|\tau_0| = \pm|\xi_0| \neq 0$ is the straight line $\Gamma = \{(t,x,\tau_0,\xi_0)\colon x = x_0 - (\xi_0/\tau_0)t\}$. Suppose that we wanted to consider the microlocal regularity of a solution to the inhomogeneous wave equation $\Box u = f(t,x)$ along Γ. It would be enough to consider the local regularity of $b_0(t,x,D)u$, where $b_0(t,x,\tau,\xi)$ is conically supported near Γ and is microlocally elliptic on a smaller conic neighborhood of Γ. The idea is a classical one: examine the equation satisfied by $b_0(t,x,D)u$, which means using a "commutator argument". Then

$$\begin{aligned}\Box b_0(t,x,D)u &= b_0(t,x,D)\Box u + [\Box, b_0(t,x,D)]u \\ &= b_0(t,x,D)f(t,x) + [\Box, b_0(t,x,D)]u,\end{aligned}$$

and the commutator is given by

$$\begin{aligned}[\Box, b_0(t,x,D)] &= 2(\partial_t b_0)(t,x,D)\partial_t - 2(\nabla_x b_0)(t,x,D)\cdot \nabla_x \\ &\quad + (\partial_t^2 b_0)(t,x,D) - (\nabla_x^2 b_0)(t,x,D).\end{aligned}$$

The symbol of the commutator operator is

$$2i(\tau\partial_t b_0 - \xi\cdot \nabla_x b_0) + (\partial_t^2 b_0 - \nabla_x^2 b_0).$$

In order for this expression to have the lowest possible order we would want $\tau\partial_t b_0 = \xi\cdot\nabla_x b_0$, that is, $b_0(t,x,\tau,\xi)$ should be constant along the bicharacteristics. Thus we let $\psi(\tau,\xi)$ be smooth, homogeneous of degree zero for $|(\tau,\xi)| \geq \varepsilon$, with conic support near (τ_0,ξ_0) and nonzero at (τ_0,ξ_0), choose $\varphi(y) \in C^\infty(\mathbf{R}^n)$ with support near 0 and nonzero at 0, and set $b_0(t,x,\tau,\xi) = \varphi(x+(\xi/\tau)t)\psi(\tau,\xi)$. Then $[\Box, b_0(t,x,D)] = a_0(t,x,D)$ is an operator of order zero. Assume that $u \in H^s_{ml}(0,x_0,\tau_0,\xi_0)$ and suppose inductively that $u \in H^{s-1}_{ml}(\Gamma)$. If u satisfies

$$\Box u = f(t,x), \text{ with } f(t,x) \in H^{s-1}_{ml}(\Gamma),$$

then the above expressions for the commutator and the boundedness of pseudodifferential operators easily imply that

$$\Box b_0(t,x,D)u \in H^{s-1}_{loc}(\mathbf{R}^{n+1}) \text{ and } b_0(t,x,D)u \in H^s_{loc}(\{|t| < \varepsilon\})$$

for some $\varepsilon > 0$. Therefore the linear energy inequality applies, yielding that $b_0(t,x,D)u \in H^s_{loc}(\mathbf{R}^{n+1})$ and hence that $u \in H^s_{ml}(\Gamma)$. The same type of argument (see Nirenberg [54]) gives the general microlocal result.

Theorem 1.8 (Hörmander). *Let* $p(x,D)$ *be a strictly hyperbolic partial* *differential operator of order* m *on* $\mathbf{R}^{n+1}$, *let* $p_m(x_0,\xi_0) = 0$, *and let* Γ *be the null bicharacteristic through* (x_0,ξ_0) *for* p_m. *If* $p(x,D)u = f(x)$, $f \in H^{s-m+1}{}_{ml}(\Gamma)$, *and* $u \in H^s{}_{ml}(x_0,\xi_0)$, *then* $u \in H^s{}_{ml}(\Gamma)$.

This result may be used to explain the properties of the singular supports of the solutions to (1.2) given at the beginning of this chapter. For instance, if u is given as in example (b), then it is easy to compute that $WF(u) \cap \{t = 0\} = \{(0,0,\tau,\xi): |\tau| = |\xi|\}$. It follows from Theorem 1.8 and computation of the null bicharacteristics that $WF(u) = \{(t,x,\tau,\xi): |t| = |x|, \xi = -(x/t)\tau, \tau \neq 0\}$, since $H^\infty{}_{ml}$ regularity is propagated along all null bicharacteristics through points $(0,x)$ with $x \neq 0$, and at any point (τ,ξ) with $|\tau| \neq |\xi|$, the microlocal ellipticity of $\Box$ implies that $u \in H^\infty{}_{ml}(t,x,\tau,\xi)$.

If we wish to extend this result to semilinear equations, for instance (1.1), a difficulty arises immediately. Even though regularity measured by differential operators is preserved under the action of nonlinear functions by the chain rule, as in Lemma 1.5, in general, microlocal regularity is not preserved. For example, we can easily construct $u \in H^s{}_{loc}(\mathbf{R}^n)$, $s > n/2$, satisfying *sing supp* $(u) = \{0\}$, $\Pi_\xi WF(u) = K_1 \cup K_2$, with K_1 a conic neighborhood of $\{(\xi_1,0,\ldots,0): \xi_1 > 0\}$ and K_2 a conic neighboorhood of $\{(0,\xi_2,\ldots,0): \xi_2 > 0\}$, and $u\hat{}(\xi) \geq 0$, $u\hat{}(\xi) \geq c\langle\xi\rangle^{-(s+n/2+\delta)}$ on smaller conic neighboorhoods of the two rays. Upon examination of the Fourier transform of u^2, it easily follows that, for K_3 a small conic neighboorhood of $\{(\rho,\rho,\ldots,0): \rho > 0\}$, $u \in H^\infty{}_{ml}(0,K_3)$, while $u^2 \notin H^\infty{}_{ml}(0,K_3)$. See Figure 1.3. It should be noted that even though the overall regularity of this example u is $H^s{}_{loc}(\mathbf{R}^n)$, on the new directions in the wave front set u^2 has better than H^s microlocal regularity. This property is true in general, and is the starting point for a theory of nonlinear microlocal analysis.

Lemma 1.9 (Rauch). *Let* $u_1, u_2 \in H^s(\mathbf{R}^n)$ *for* $s > n/2$, *and suppose that* $\Pi_\xi WF(u_i) \subset K_i$ *for* $i = 1, 2$. *If* $K_3 \subset (K_1 \cup K_2)^{comp}$, *then for any* $\delta > 0$, $u_1 u_2 \in H^{2s-n/2-\delta}{}_{ml}(\mathbf{R}^n \times K_3)$.

Proof. It is enough to assume that $supp(u_i\hat{}) \subset K_i$, since the remaining terms are smooth and are easily handled. Then

$$\chi_{K3}\langle\xi\rangle^r (u_1 u_2)\hat{}(\xi) = \int K(\xi,\eta) f\hat{}(\xi-\eta) g\hat{}(\eta)\,d\eta\,,$$

with $f, g \in L^2(\mathbf{R}^n)$, and

$$K(\xi,\eta) = \langle\xi\rangle^{r}\langle\xi-\eta\rangle^{-s}\langle\eta\rangle^{-s}\chi_{K3}(\xi)\chi_{K1}(\xi-\eta)\chi_{K2}(\eta).$$

If $\xi \in K_3$, $\eta \in K_2$, and $|\xi-\eta| \ll |\xi|$, then $\eta = [(\eta-\xi)+\xi] \in K_3$, which does not occur on the support of the integrand. Hence $|\xi-\eta| \geq \varepsilon|\xi|$ on $supp(K)$. Similarly, $\xi \in K_3$ and $(\xi-\eta) \in K_1$ imply that $|\eta| \geq \varepsilon|\xi|$ on $supp(K)$. Thus if $r < 2s - n/2$, it follows that $K(\xi,\eta) \leq C\langle\xi\rangle^{-n/2-\delta}$. Consequently, by Lemma 1.4 the required estimate $\chi_{K3}\langle\xi\rangle^{r}(u_1u_2)^{\wedge}(\xi) \in L^2(\mathbf{R}^n)$ holds. Q.E.D.

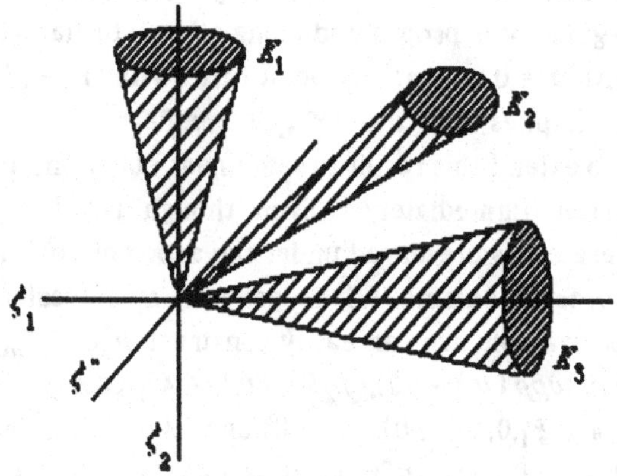

Figure 1.3

Corollary 1.10. *Let Γ be a closed conic subset of $T^*(\mathbf{R}^n)\backslash 0$. Then $H^s_{loc}(\mathbf{R}^n) \cap H^r_{ml}(\Gamma)$ is an algebra for $n/2 < s \leq r < 2s - n/2$.*

Proof. It suffices to assume that u and v have support near x_0 and to prove that $uv \in H^r_{ml}(\Gamma)$ for $u, v \in H^s_{loc}(\mathbf{R}^n) \cap H^r_{ml}(\Gamma)$. For K a small conic neighborhood of $\Pi_\xi\Gamma$, we write $u = u_1 + u_2$, with

$$u_1^{\wedge}(\xi) + u_2^{\wedge}(\xi) = \chi_K u^{\wedge}(\xi) + \chi_{K\,comp}\, u^{\wedge}(\xi),$$

and similarly $v = v_1 + v_2$. Then $u_1v_1 \in H^r_{loc}(\mathbf{R}^n)$, and from Lemma 1.9, $u_2v_2 \in H^{2s-n/2-\delta}_{ml}(\Gamma)$. In addition, the proof of Lemma 1.9 easily shows that the remaining terms satisfy $u_1v_2 \in H^r_{ml}(\Gamma)$ and $u_2v_1 \in H^r_{ml}(\Gamma)$.
Q.E.D.

The first proof of these results was given in Rauch [57]. More generally, the analogue of (1.5) holds for the product of functions of differing regularities. Bony [15] established that $H^s_{loc}(\mathbf{R}^n) \cap H^r_{ml}(G)$ is preserved for $n/2 < s \leq r < 2s - n/2$ under the action of smooth functions $f(u)$ by introducing the "paraproduct" of nonsmooth functions, and Meyer [52] extended this property to $r \leq 2s - n/2$. For a proof of the invariance under smooth functions for $r \leq 2s - n/2$ along the lines of the proof of Lemma 1.5, see Beals [7].

As an application, we have the following microlocal regularity result of Rauch [57] for solutions to semilinear wave equations.

Theorem 1.11. *Let $p(x,D)$ be a strictly hyperbolic partial differential operator of order m on $\mathbf{R}^{n+1}$, let $p_m(x_0,\xi_0) = 0$, and let Γ be the null bicharacteristic through (x_0,ξ_0) for p_m. Suppose that $u \in H^s_{loc}(\mathbf{R}^{n+1})$, $s > (n+1)/2 + m - 2$, f is smooth, and $p(x,D)u = f(x,u,\ldots,D^{m-2}u)$. If $u \in H^r_{ml}(x_0,\xi_0)$, then $u \in H^r_{ml}(\Gamma)$ for $r < 2s - (n+1)/2 - m + 3$.*

Such solutions exist locally by the usual contraction mapping argument, using Schauder's Lemma. The global hypotheses may be replaced by appropriate local ones, using finite propagation speed for the strictly hyperbolic operator.

Proof. Suppose that $u \in H^t_{ml}(\Gamma)$ for $s \leq t < 2s - (n+1)/2 - m + 2$. By the algebra property, since $(t - m + 2) < 2(s - m + 2) - (n+1)/2$,

$$f(x,u,\ldots,D^{m-2}u) \in H^{s-m+2}_{loc}(\mathbf{R}^{n+1}) \cap H^{t-m+2}_{ml}(\Gamma),$$

and if $t \leq r - 1$, $u \in H^{t+1}_{ml}(x_0,\xi_0)$. Hörmander's Theorem then implies that $u \in H^{t+1}_{ml}(\Gamma)$. The result follows by induction on t. Q.E.D.

The analogue of this nonlinear microlocal propagation of regularity result holds for more general strictly hyperbolic equations of the form (0.1), (0.2), and (0.3). Bony [15] originally obtained these estimates using the general theory of paradifferential operators. The simple argument given above does not extend directly, for example to the equation $\Box u = f(t,x,u,Du)$, since the analogue of the proof of Hörmander's Theorem would be needed, and the pseudodifferential cutoff operator $b_0(t,x,D)$ would have to be commuted past a nonlinear function of Du. Moreover, for the quasilinear equation the corresponding operator $b_0(t,x,D)$ would no longer have smooth coefficients, since those coefficients would now depend on u and its derivatives. The argument above can be extended by expanding the pseudodifferential calculus to such operators. The full calculus cannot be expected to extend, since for example in the smooth case if $p(D)$ is pseudodifferential and not differential, and A stands for the operator given by multiplication by $a(x)$, the composition has an infinite asymptotic expansion in terms of operators of decreasing order,

$$p(D)A \sim \Sigma(\alpha!)^{-1}(D_x^\alpha a)(x)(\partial_\xi^\alpha p)(\xi).$$

If $a(x)$ is no longer assumed to be smooth, multiplication by $D_x^\alpha a(x)$ will not preserve the Sobolev space $H^s_{loc}(\mathbf{R}^n)$ if α is large enough. But an approximate calculus can be defined, either through the use of paradifferential operators or as in Beals-Reed [13], [14], Chen [25] by treating operators with $H^s_{loc}(\mathbf{R}^n)$ or $H^s_{loc}(\mathbf{R}^n) \cap H^r_{ml}(G)$ coefficients, breaking off the expansions of compositions after finitely many steps, and estimating the smoothing properties of the remainders.

As an example of the application of the calculus developed in [13], [14], consider a solution $u \in H^s_{loc}(\mathbf{R}^{n+1})$ to

$$\Box u = f(t,x,u,Du). \tag{1.6}$$

Here f is assumed to be a smooth function of its arguments, and Du represents the vector $(\partial_t u, \nabla_x u)$. If v stands for the vector of all second derivatives of u, then differentiation of (1.6) yields a vector equation of the form

$$\Box v = f_1(t,x,u,Du)Dv + f_0(t,x,u,Du,v). \tag{1.7}$$

The vector valued functions f_1 and f_0 are smooth, and we assume that $s' = s - 2 > (n+1)/2$ so that the applications of the chain rule are justified.

Let Γ be the null bicharacteristic through $(0,x_0,\tau_0,\xi_0)$ for $\Box$, and assume inductively that $u \in H^r_{ml}(\Gamma)$ for $r < 2s - 2 - (n+1)/2$. Then for $r' = r - 2$, we have $v \in H^{s'}_{loc}(\mathbf{R}^{n+1}) \cap H^{r'}_{ml}(\Gamma)$ and $r' < 2s' - (n+1)/2$. Thus by Corollary 1.10,

$$f_0(t,x,u,Du,v) \in H^{s'}_{loc}(\mathbf{R}^{n+1}) \cap H^{r'}_{ml}(\Gamma),$$
$$f_1(t,x,u,Du) \in H^{s'+1}_{loc}(\mathbf{R}^{n+1}) \cap H^{r'+1}_{ml}(\Gamma).$$

If $u \in H^{r+1}{}_{ml}(0,x_0,\tau_0,\xi_0)$, we would like to mimic the proof of Theorem 1.8 to conclude that $u \in H^{r+1}{}_{ml}(\Gamma)$, or, equivalently, $v \in H^{r'+1}{}_{ml}(\Gamma)$. Let $b_0(t,x,\tau,\xi)$ be constructed as before, so that $[\Box, b_0(t,x,D)] = p_0(t,x,D)$ has order zero and is supported on a small conic neighborhood of Γ. Then from (1.6),

$$\Box b_0(t,x,D)v = b_0(t,x,D)\{f_1(t,x,u,Du)Dv\} \\ + b_0(t,x,D)f_0(t,x,u,Du,v) + p_0(t,x,D)v.$$

If the support of b_0 is sufficiently small, then

$$b_0(t,x,D)f_0(t,x,u,Du,v) + p_0(t,x,D)v \in H^{r'}{}_{loc}(\mathbf{R}^{n+1}).$$

Moreover, if $b'_0(t,x,\tau,\xi)$ has small conic support near Γ and is identically one on the support of b_0, then modulo a smooth remainder,

$$b_0(t,x,D)\{f_1 Dv\} = b'_0(t,x,D)b_0(t,x,D)\{f_1 Dv\} \\ = b'_0(t,x,D)\{f_1 D b_0(t,x,D)v\} + b'_0(t,x,D)\{f_1[b_0(t,x,D),D]v\} \\ + b'_0(t,x,D)[b_0(t,x,D),f_1]Dv.$$

Since $[b_0(t,x,D),D]v \in H^{s'}{}_{loc}(\mathbf{R}^{n+1}) \cap H^{r'}{}_{ml}(\Gamma)$, it follows from Corollary 1.10 that $b'_0(t,x,D)\{f_1[b_0(t,x,D),D]v\} \in H^{r'}{}_{loc}(\mathbf{R}^{n+1})$. Suppose it can be established that

(1.8) $\quad [b_0(t,x,D),f_1]$ maps $H^{s'-1}{}_{loc}(\mathbf{R}^{n+1}) \cap H^{r'-1}{}_{ml}(\Gamma)$ into $H^{s'}{}_{loc}(\mathbf{R}^{n+1}) \cap H^{r'}{}_{ml}(G)$.

Then, with $w = b_0(t,x,D)v$, it follows that

$$\Box w = b'_0(t,x,D)\{f_1 Dw\} + g_1,$$

(1.9) $\quad g_1 \in H^{r'}{}_{loc}(\mathbf{R}^{n+1})$, $f_1 \in H^{s'+1}{}_{loc}(\mathbf{R}^{n+1}) \cap H^{r'+1}{}_{ml}(\Gamma)$, $w \in H^{r'+1}{}_{loc}((0,x_0))$, and b'_0 has small conic support near G.

The linear energy inequality implies that $w \in H^{r'+1}{}_{loc}(\mathbf{R}^{n+1})$, as long as the norm of $b'_0(t,x,D)\{f_1 Dw\}$ in $H^{r'}(\mathbf{R}^{n+1})$ can be estimated in terms of the norms of Dw in $H^{r'}(\mathbf{R}^{n+1})$ and of f_1 in $H^{s'+1}(\mathbf{R}^{n+1}) \cap H^{r'+1}{}_{ml}(G)$. The proof of Corollary 1.10 yields exactly such an estimate. By induction on r', it follows that if $v \in H^{r'+1}{}_{ml}(0,x_0,\tau_0,\xi_0)$ for $r' < 2s' - (n+1)/2$, then $v \in$

$H^{r+1}_{ml}(\Gamma)$. Hence if $u \in H^r_{ml}(0, x_0, \tau_0, \xi_0)$ for $r < 2s - (n+1)/2 - 1$, then $u \in H^r_{ml}(\Gamma)$. The same argument works if $\Box$ is replaced by any second order strictly hyperbolic operator. Therefore, subject to the proof of (1.8), the following microlocal regularity result for the general semilinear equation of second order is established. (The statement for the equation (0.1) of order m is analogous.)

Theorem 1.12. *Let $p_2(x,D)$ be a second order strictly hyperbolic partial differential operator on $\mathbf{R}^{n+1}$, let $p_2(x_0,\xi_0) = 0$, and let G be the null bicharacteristic through (x_0,ξ_0) for p_2. Suppose that $u \in H^s_{loc}(\mathbf{R}^{n+1})$, $s > (n+1)/2 + 2$, f is smooth, and $p_2(x,D)u = f(x,u,Du)$. If $u \in H^r_{ml}(x_0,\xi_0)$, then $u \in H^r_{ml}(\Gamma)$ for $r < 2s - (n+1)/2 - 1$.*

The key estimate (1.8) follows immediately from the commutator result established in [13], which may be extended to a more general calculus.

Lemma 1.13. *Suppose that $m(x) \in H^s_{loc}(\mathbf{R}^n) \cap H^r_{ml}(\Gamma)$ and $u(x) \in H^{s-1}_{loc}(\mathbf{R}^n) \cap H^{r-1}_{ml}(\Gamma)$, for $n/2 < s \le r < 2s - n/2$. If $b_0(x,\xi) \in S^0_{1,0}$ has conic support sufficiently near Γ, then $[b_0(x,D), m(x)]u \in H^r_{ml}(\Gamma)$.*

Proof. For notational convenience, we consider the constant coefficient case $b_0 = b_0(\xi)$; the general case is similar. Assume that u and m have been multiplied by smooth functions with compact support near x_0. Let $G \subset G'$ be small conic neighborhoods of $\{\xi\colon (x_0,\xi) \in \Gamma\}$, and let $B = \mathbf{R}^n \setminus G'$. Then

$$
\begin{aligned}
[b_0(D), m(x)]u\hat{\ }(\xi) &= b_0(\xi)(mu)\hat{\ }(\xi) - \int m\hat{\ }(\xi-\eta)\, b_0(\eta)\, u\hat{\ }(\eta)\, d\eta \\
&= \int (b_0(\xi) - b_0(\eta))\, m\hat{\ }(\xi-\eta)\, u\hat{\ }(\eta)\, d\eta\,.
\end{aligned}
$$

From the assumptions on m and u it therefore suffices to show that

$$
\int K_i(\xi,\eta)\, f\hat{\ }(\xi-\eta)\, g\hat{\ }(\eta)\, d\eta \in L^2(\mathbf{R}^n)
$$

for $f, g \in L^2(\mathbf{R}^n)$, with

$$
\begin{aligned}
K_1(\xi,\eta) &= \chi_G(\xi)\chi_B(\xi-\eta)\chi_B(\eta)(b_0(\xi) - b_0(\eta))\langle\xi\rangle^r / \langle\xi-\eta\rangle^{s+1}\langle\eta\rangle^{s-1}, \\
K_2(\xi,\eta) &= \chi_G(\xi)\chi_G(\xi-\eta)\chi_B(\eta)(b_0(\xi) - b_0(\eta))\langle\xi\rangle^r / \langle\xi-\eta\rangle^{r+1}\langle\eta\rangle^{s-1}, \\
K_3(\xi,\eta) &= \chi_G(\xi)\chi_G(\xi-\eta)\chi_G(\eta)(b_0(\xi) - b_0(\eta))\langle\xi\rangle^r / \langle\xi-\eta\rangle^{r+1}\langle\eta\rangle^{r-1}, \\
K_4(\xi,\eta) &= \chi_G(\xi)\chi_B(\xi-\eta)\chi_G(\eta)(b_0(\xi) - b_0(\eta))\langle\xi\rangle^r / \langle\xi-\eta\rangle^{s+1}\langle\eta\rangle^{r-1}.
\end{aligned}
$$

Since b_0 is bounded, the proofs of Lemma 1.9 and Corollary 1.10 yield that $|K_1| \le C/\langle \xi \rangle^{2s-r}$ and $|K_2| \le C/\langle \xi - \eta \rangle \langle \eta \rangle^{s-1}$. For the remaining kernels, where $|\xi - \eta| \ge \epsilon|\xi|$ it follows that $|K_3| \le C/\langle \xi - \eta \rangle \langle \eta \rangle^{r-1}$ and (since $|\eta| \ge \epsilon|\xi|$ on the support of K_4) $|K_4| \le C/\langle \xi \rangle^s$. On the other hand, if $|\xi - \eta| \le \epsilon|\xi|$, it follows that η and ξ are comparable, so $\eta + t(\xi - \eta)$ is comparable to η for $0 \le t \le 1$. Then, since $\nabla b_0 \in S^{-1}_{1,0}$ and

$$b_0(\xi) - b_0(\eta) = (\xi - \eta)\cdot \int_0^1 \nabla b_0(\eta + t(\xi - \eta))\,dt,$$

it follows that for such values of η and ξ, $|b_0(\xi) - b_0(\eta)| \le C\langle \xi - \eta \rangle / \langle \eta \rangle$. Therefore, for $|\xi - \eta| \le \epsilon|\xi|$, $|K_3| \le C/\langle \xi - \eta \rangle^r$ and $|K_4| \le C/\langle \xi - \eta \rangle^s$. The desired estimates now follow from Lemma 1.4. Q.E.D.

As an example of the application of Bony's paradifferential calculus, we consider microlocal elliptic regularity for solutions of the general fully nonlinear equation (0.3). Following Meyer [52], we recall some properties of the Littlewood-Paley decomposition of a function $v \in H^s(\mathbf{R}^n)$. Since we are interested in local regularity, by the Paley-Wiener Theorem it may be assumed that all of the functions under consideration have Fourier spectrum supported where $|\xi| \ge 2$. Let $\varphi \in C^\infty(\mathbf{R}^n)$ be radial, have support in $1 \le |\xi| \le 3$, and satisfy

$$\sum_{k=1}^{\infty} \varphi_k(\xi) = 1 \quad \text{for} \quad |\xi| \ge 2\text{, with } \varphi_k(\xi) = \varphi(|\xi|/2^k).$$

Let Δ_k denote the zero order pseudodifferential operator $\varphi_k(D)$. Then

$$v = \sum_{k=1}^{\infty} \Delta_k v,$$

and $(\Delta_k v)$ is called the Littlewood-Paley decomposition of v. The Fourier transform of $\Delta_k v$ is supported where $|\xi|$ is comparable to 2^k. As is easily verified, $\|v\|_{H^s}$ is equivalent to $[\sum 2^{2ks}(\|\Delta_k v\|_{L^2})^2]^{1/2}$. In particular, if $v \in H^s(\mathbf{R}^n)$, then $\|\Delta_k v\|_{L^2} \le C2^{-2ks}$, so that

$$\begin{aligned}\|\Delta_k v\|_{L^\infty} &\le \int_{|\xi| \approx 2^k} |\Delta_k v\hat{\ }(\xi)|\,d\xi \\ &\le C2^{kn/2}\|\Delta_k v\hat{\ }\|_{L^2} = C2^{kn/2}\|\Delta_k v\|_{L^2} \\ &\le C\,2^{-k(s-n/2)}.\end{aligned}$$

Therefore

$$\|\Delta_k v\|_{L^\infty} \leq C 2^{-k(s-n/2)} \|v\|_{H^s}. \tag{1.10}$$

Set $S_k = 0$ for $k \leq 0$ and $S_k = \Delta_1 + \ldots + \Delta_k$ for $k \geq 1$, so that $\Delta_k = S_k - S_{k-1}$. We want to consider the relationship between the Littlewood-Paley decomposition and a nonlinear function of v such as v^2. If $v \in H^s(\mathbf{R}^n)$ for $s > n/2$, then v is the pointwise limit of $S_k v$ as $k \to \infty$, and v^2 is the pointwise limit of $(S_k v)^2$. In particular, we have

$$\begin{aligned} v^2 &= \sum (S_k v)^2 - (S_{k-1} v)^2 = \sum (S_k v + S_{k-1} v) \Delta_k v \\ &= \sum 2(S_{k-1} v) \Delta_k v + \sum (D_k v)^2 \\ &= \sum 2(S_{k-3} v) \Delta_k v + \sum (2\Delta_{k-1} v + \Delta_k v)(\Delta_k v) = p(x,D) v + w. \end{aligned}$$

Here $p(x,D)$ is an operator of order zero with symbol

$$p(x,\xi) = \sum 2(S_{k-3} v) \varphi_k(\xi).$$

The key idea is to note that the remainder term w is smoother than v if $v \in H^s(\mathbf{R}^n)$ for $s > n/2$, while in addition $p(x,\xi)$ is in a sufficiently good symbol class to allow for a symbolic calculus. Since the terms in the expression for w are quadratic in the Littlewood-Paley decomposition for v, the extra smoothness of this remainder is unsurprising.

For the general nonlinear function f and v as above, we write

$$\begin{aligned} f(v) &= \sum f(S_k v) - f(S_{k-1} v) \\ &= (S_k v) - f(S_{k-1} v) = f(S_{k-1} v + \Delta_k v) - f(S_{k-1} v) \\ &= \left[\int_0^1 f'(S_{k-1} v + t\Delta_k)\, dt \right] \Delta_k v. \end{aligned}$$

We imitate the above analysis and let $p(x,D) = p(x,f,v,D)$ be the pseudo- differential operator with symbol

$$p(x,\xi) = \sum S_{k-3} f'(v) \varphi_k(\xi). \tag{1.11}$$

We recall that the class $S^m_{1,1}$ is defined to be the collection of symbols $a(x,\xi)$ satisfying $|D_x^\alpha D_\xi^\beta a(x,\xi)| \leq C_{\alpha,\beta} \langle \xi \rangle^{m + |\alpha| - |\beta|}$ for all multiindices.

Lemma 1.14. *Let* $v \in H^s(\mathbf{R}^n)$, $s > n/2$, *let* $f \in C^\infty(\mathbf{R}^n)$, *and suppose that* $p(x,D) = p(x,f,v,D)$ *has symbol given by* (1.11). *Then there is a symbol* $q(x,\xi) \in S^{-(s-n/2)}_{1,1}$ *such that* $f(v) - p(x,D)v = q(x,D)v$.

Proof. From above,

$$q(x,\xi) = \sum_{k=1}^{\infty}\left[\int_0^1 f'(S_{k-1}v + t\Delta_k v)\,dt - S_{k-3}f'(v)\right]\varphi_k(\xi).$$

First, we write the term in parentheses as

$$\int_0^1 f'(S_{k-1}v + t\Delta_k v) - f'(S_{k-1}v + t\Delta_k + (1-t)\Delta_k v + (I-S_k)v)\,dt + (I - S_{k-3})f'(v)$$

$$= \int_0^1\left[\int_0^1 f''(S_{k-1}v + t\Delta_k v + r(1-t)\Delta_k v + r(I-S_k)v)\,dr\right]\left[(1-t)\Delta_k v + (I-S_k)v\right]dt$$

$$+ (I - S_{k-3})f'(v).$$

Now, $S_k v$ and $\Delta_k v$ are in $H^s(\mathbf{R}^n) \subset L^\infty(\mathbf{R}^n)$ since $s > n/2$, so f' is bounded. Moreover, we can asume without loss of generality that f, which is allowed to depend also on x, has compact support in x, so that $v, f'(v) \in H^s(\mathbf{R}^n)$. Therefore, by (1.10),

$$\|(I - S_k)v\|_{L^\infty} \le \sum_{j>k}\|\Delta_k v\|_{L^\infty} \le C2^{-k(s-n/2)},$$

and similarly $\|(I - S_{k-3})f'(v)\|_{L^\infty} \le C2^{-k(s-n/2)}$. Since $|\xi| \approx 2^k$ on supp(φ_k), it follows that $|q(x,\xi)| \le C\langle\xi\rangle^{-(s-n/2)}$.

Next, if $|\alpha| > n/2$, there are smooth functions g_β for $\beta_1 + \ldots + \beta_m = \alpha$ such that

$$\partial_x^\alpha f'(S_{k-1}v + t\Delta_k v) = \sum g_\beta(S_{k-1}v + t\Delta_k v)\,\partial_x^{\beta_1}(S_{k-1}v + t\Delta_k v)\cdots\partial_x^{\beta_m}(S_{k-1}v + t\Delta_k v).$$

Since

$$\begin{aligned}\|\partial^\beta(S_{k-1}v + t\Delta_k v)\|_{L^\infty} &\le \sum_{j\le k}\int_{|\xi|\approx 2^j}|\partial^\beta\Delta_j v\hat{}(\xi)|\,d\xi\\ &\le C\sum_{j\square k}2^{jn/2}\|\partial^\beta\Delta_j v\|_{L^2} \le C\sum_{j\le k}2^{j(-(s-n/2)+|\beta|)}\\ &\le C(1 + 2^{k(-(s-n/2)+|\beta|)}),\end{aligned}$$

it follows from the above expression that for $|\alpha| > s - n/2$,

$$\|\partial_x^\alpha f'(S_{k-1}v + t\Delta_k v)\|_{L^\infty} \leq C2^{k(-(s-n/2)+|\alpha|)}.$$

Similarly,

$$\|\partial_x^\alpha S_{k-3}f'(v)\|_{L^\infty} \leq C2^{k(-(s-n/2)+|\alpha|)}.$$

Therefore, from above,

$$|\partial_x^\alpha q(x,\xi)| \leq C\langle\xi\rangle^{-(s-n/2)+|\alpha|}$$

for $\alpha = 0$ and for $|\alpha| > s - n/2$, and so for all α by interpolation. The estimate $|\partial_x^\alpha \partial_\xi^\beta q(x,\xi)| \leq C\langle\xi\rangle^{-(s-n/2)+|\alpha|-|\beta|}$ follows in exactly the same fashion. Q.E.D.

Operators in the symbol class $S^{-r}_{1,1}$ are not smoothing on Sobolev spaces of all orders, but do map $H^s(\mathbf{R}^n)$ to $H^{s+r}(\mathbf{R}^n)$ as long as $s + r > n/2$. (See Coifman-Meyer [29].) Consequently, the remainder term in the expression for $f(v)$ above is strictly smoother than v.

Corollary 1.15. *Under the above hypotheses,* $f(v) = p(x,D)v + w$, *with* $w \in H^{2s-n/2}(\mathbf{R}^n)$.

If $p(x,\xi) = \Sigma S_{k-3}f'(x)\varphi_k(\xi)$ with $f \in H^s(\mathbf{R}^n)$, $s > n/2$, the proof of Lemma 1.14 yields that $p(x,\xi) \in S^0_{1,1}$. In general, operators in this class do not have a good symbolic calculus. In this case, though, the symbol of p has partial Fourier transform $p^\wedge(\eta,\xi)$ with the special property that $|\eta| \leq |\xi|/2$ on its support. Moreover, for each fixed ξ, at most three of the terms $\varphi_k(\xi)$ are non-zero, so that the smoothness of $p(x,\xi)$ in x is equivalent to that of f. Precisely,

$$\|D_\xi^\beta p(x,\xi)\|_{H^s} \leq 3\,sup_k\|S_{k-3}f'\|_{H^s}\langle\xi\rangle^{-|\beta|} \leq C_\beta\|f'\|_{H^s}\langle\xi\rangle^{-|\beta|}.$$

Such symbols have a good partial calculus for composition with elements of $S^0_{1,1}$. The corresponding operators are called paradifferential operators.

Lemma 1.16. *Let $p(x,\xi)$ have partial Fourier transform $p\hat{}(\eta,\xi)$ with support where $|\eta| \le |\xi|/2$, and suppose that for all β, $\|D_\xi^\beta p(x,\xi)\|_{H^s} \le C_\beta \|f\|_{H^s} \langle \xi \rangle^{-|\beta|}$ for $s > n/2$. Let $q(x,\xi) \in S^0_{1,1}$ and let k be the greatest integer in $s - n/2$. Then*

$$q(x,D)p(x,D) - \sum_{|\alpha| \le k} \frac{1}{\alpha!}(\partial_\xi^\alpha q\, D_x^\alpha p)(x,D) = r(x,D),$$

with $r(x,\xi) \in S^{-(s-n/2)}_{1,1}$.

Proof. For $u \in C^\infty_{com}(\mathbf{R}^n)$,

$$\begin{aligned} q(x,D)p(x,D)u(x) &= \int e^{ix\cdot\xi} q(x,\eta)p\hat{}(\eta-\xi,\xi)u\hat{}(\xi)\,d\xi d\eta \\ &= \int e^{ix\cdot\xi}\int e^{ix\cdot(\eta-\xi)} q(x,\eta)p\hat{}(\eta-\xi,\xi)\,d\eta\, u\hat{}(\xi)\,d\xi \\ &= \int e^{ix\cdot\xi}\int e^{ix\cdot\eta} q(x,\eta+\xi)p\hat{}(\eta,\xi)\,d\eta\, u\hat{}(\xi)\,d\xi \\ &= \int e^{ix\cdot\xi}\Big(\int e^{ix\cdot\eta}\textstyle\sum_{|\alpha|\le k}(1/\alpha!)\,\partial_\xi^\alpha q(x,\xi)\,\eta^\alpha p\hat{}(\eta,\xi)\,d\eta + r(x,\xi)\Big)u\hat{}(\xi)\,d\xi \\ &= \textstyle\sum_{|\alpha|\le k}(1/\alpha!)\,\partial_\xi^\alpha q\, D_x^\alpha p(x,D)u(x) + r(x,D)u(x), \end{aligned}$$

with

$$r(x,\xi) = \sum_{|\alpha| = k+1} c_\alpha \int e^{ix\cdot\eta}\Big[\int_0^1 \partial_\xi^\alpha q(x,\xi+t\eta)\,dt\Big]\eta^\alpha p\hat{}(\eta,\xi)\,d\eta.$$

Since $\xi + t\eta$ is comparable to ξ on the support of $p\hat{}(\eta,\xi)$ for $0 \le t \le 1$, and since $q(x,\xi) \in S^0_{1,1}$, it follows that

$$|\partial_\xi^\alpha q(x,\xi+t\eta)\,\eta^\alpha p\hat{}(\eta,\xi)| \le C\frac{\langle\eta\rangle^{k+1}}{\langle\xi\rangle^{k+1}}|p\hat{}(\eta,\xi)|.$$

The norm estimates on p yield a function $f \in L^2(\mathbf{R}^n)$ with $|p\hat{}(\eta,\xi)| \le C|f(\eta)|/\langle\xi\rangle^s$. Hence

$$|r(x,\xi)| \le C\int_{|\eta|\le|\xi|/2} \frac{\langle\eta\rangle^{k+1}}{\langle\xi\rangle^{k+1}}\frac{|p\hat{}(\eta,\xi)|}{\langle\eta\rangle^s}\,d\eta \le C\frac{|f(\eta)|}{\langle\eta\rangle^s}$$

$$\le \frac{C}{\langle\xi\rangle^{k+1}}\Big[\int_{|\eta|\le|\xi|/2}\langle\eta\rangle^{2(k+1-s)}\,d\eta\Big]^{1/2} \le C\frac{\langle\xi\rangle^{k+1-s+n/2}}{\langle\xi\rangle^{k+1}} = C\langle\xi\rangle^{-(s-n/2)},$$

since $k+1 > s - n/2$. The estimates on $|D_x^\alpha D_\xi^\beta r(x,\xi)|$ follow in exactly the same fashion from the estimates on $|D_x^\alpha D_\xi^\beta q(x,\xi+t\eta)|$. Q.E.D.

This calculus may be applied in the usual fashion to deduce microlocal elliptic regularity for solutions to nonlinear equations which have been linearized using paradifferential operators.

Theorem 1.17. *Let* $f(x, y, \ldots, y_\alpha)_{|\alpha| \le m}$ *be a smooth function, and let* $u \in H^s_{loc}(\mathbf{R}^n)$, $s > n/2 + m$ *satisfy* $f(x, u, \ldots, D^\alpha u) = 0$. *Suppose that*

$$\sum_{|\alpha| = m} \frac{\partial f}{\partial y^\alpha}(x_0, u(x_0), \ldots, D^\alpha u(x_0))\xi_0^\alpha \neq 0 .$$

Then $u \in H^{2s-n/2-m}(x_0, \xi_0)$.

Proof. Let v denote the vector $(u, \ldots, D^\alpha u)$, so that $v \in H^{s-m}_{loc}(\mathbf{R}^n)$ and $f(x, v) = 0$. Let $p(x, \xi) = p(f, v, x, \xi)$ be the (vector valued) symbol defined as in (1.11). By Corollary 1.15, $p(x, D) \cdot v \in H^{2(s-m-n/2)}_{loc}(\mathbf{R}^n)$. If $p_0(x, \xi) = p(x, \xi) \cdot \xi^\alpha \langle \xi \rangle^{-m/2}$, then $p_0(x, D)$ is a paradifferential operator of order zero, $p_0(x, D)(1 - \Delta)^{m/2} u \in H^{2(s-m-n/2)}_{loc}(\mathbf{R}^n)$, and by assumption, $p_0(x_0, \xi_0) \neq 0$. Let $e(x, \xi) \in S^0_{1,0}$ have small conic support and be microlocally elliptic near (x_0, ξ_0), and set $q_0(x, \xi) = e(x, \xi)/p_0(x, \xi)$. Then $q_0(x, \xi) \in S^0_{1,1}$, and by Lemma 1.16,

$$q_0(x, D)p_0(x, D) = e(x, D) + \Sigma_{1 \le |\alpha| \le k}(1/\alpha!)\, \partial_\xi^\alpha q_0 D_x^\alpha p_0(x, D) + r_0(x, D),$$

with $r_0(x, \xi) \in S^{-(s-m-n/2)}_{1,1}$. For $|\alpha| \ge 1$, $\partial_\xi^\alpha q_0 D_x^\alpha p_0(x, \xi) \in S^{-1}_{1,1}$, so we can set

$$q_{-1}(x, \xi) = -(\Sigma_{1 \le |\alpha| \le k}(1/\alpha!)\, \partial_\xi^\alpha q_0 D_x^\alpha p_0(x, \xi))/p_0(x, \xi)$$

and have

$$(q_0(x, D) + q_{-1}(x, D))p_0(x, D) = e(x, D) + \Sigma_{1 \le |\alpha| \le k}(1/\alpha!)\, \partial_\xi^\alpha q_{-1} D_x^\alpha p_0(x, D) + r_{-1}(x, D),$$

with $r_{-1}(x, \xi) \in S^{-(s-m-n/2)}_{1,1}$. This time the terms in the sum are in $S^{-2}_{1,1}$. Proceeding by induction, we obtain $q(x, \xi) \in S^0_{1,1}$ with

$$q(x, D)p(x, D) = e(x, D) + r(x, D),$$

for $r(x, \xi) \in S^{-(s-m-n/2)}_{1,1}$. Hence

$$e(x,D)(1-\Delta)^{m/2}u = q(x,D)w + r(x,D)(1-\Delta)^{m/2}u.$$

Since $q(x,D)$ is bounded on $H^{2(s-m-n/2)}(\mathbf{R}^n)$, and $r(x,D)$ maps $H^s(\mathbf{R}^n)$ to $H^{2(s-m-n/2)}(\mathbf{R}^n)$, it follows that $e(x,D)(1-\Delta)^{m/2}u \in H^{2(s-m-n/2)}{}_{loc}(\mathbf{R}^n)$, that is, $u \in H^{2s-n/2-m}{}_{ml}(x_0,\xi_0)$. Q.E.D.

For elliptic regularity, it is natural to consider the more general Sobolev spaces $L^{p,s}{}_{loc}(\mathbf{R}^n)$ corresponding to functions with derivatives up to order s in $L^p{}_{loc}(\mathbf{R}^n)$, $1 < p < \infty$, and the Lipschitz spaces $C^s{}_{loc}(\mathbf{R}^n)$. Analogous nonlinear microlocal estimates are established in Meyer [52], through the characterization of the functions in these spaces in terms of the their Littlewood-Paley decompositions.

For the propagation of regularity along null bicharacteristics, Bony [15] uses the calculus of Lemma 1.16, while in Beals-Reed [14] the extension of Lemma 1.13 is applied. Hörmander's energy estimate proof of Theorem 1.8, which is somewhat less simple than the argument presented above, can be adapted in a straightforward manner to the paradifferential or nonsmooth pseudodifferential calculus. The results for the general quasilinear equation (0.2) are stated here. The fully nonlinear equation (0.3) may be differentiated once and treated as a quasilinear equation for Du, under the hypothesis of additional smoothness for u.

Theorem 1.18. *Let $u \in H^s{}_{loc}(\mathbf{R}^{n+1})$, $s > (n+1)/2 + m$. Suppose that*

$$p_m(x,D) = \sum_{|\alpha|=m} a_\alpha(x,u,\ldots,D^{m-1}u)D^\alpha$$

is a strictly hyperbolic partial differential operator on $\mathbf{R}^{n+1}$. Let $p_m(x_0,\xi_0) = 0$, let Γ be a null bicharacteristic through (x_0,ξ_0) for p_m, and let f be smooth. If $p_m(x,D)u = f(x,u,\ldots,D^{m-1}u)$, and $u \in H^r{}_{ml}(x_0,\xi_0)$, then $u \in H^r{}_{ml}(\Gamma)$ for $r < 2s-(n+1)/2-m$.

As in the linear case, the microlocal elliptic and hyperbolic regularity results may be used to draw conclusions about the local regularity of a solution. For example, let $u \in H^s{}_{loc}(\mathbf{R}^{n+1})$, $s > (n+1)/2 + m$, satisfy the quasilinear equation (0.2), which is assumed to be strictly hyperbolic with respect to planes $\{t = c\}$. Let C denote the union of the x projections of the null bicharacteristics over the origin. (C is the analogue of the surface of the light cone over the origin for the operator $\Box$.) If $u(x,0)$, $u_t(x,0) \in C^\infty{}_{loc}(\mathbf{R}^n)\backslash 0$, then, by finite propagation speed, $u \in C^\infty_{loc}(\mathbf{R}^{n+1})$ outside of the region enclosed by C. Moreover, if $(x_0, t_0) \in \mathbf{R}^{n+1}\setminus C$, then every null bicharacteristic Γ through a hyperbolic direction (ξ_0, τ_0) over (x_0, t_0) contains points over $\{t = 0\}$ where $u \in H^\infty_{ml}$. By Theorem 1.18,

$$u \in H^{2s-(n+1)/2-m}_{ml}(x_0, t_0, \xi_0, \tau_0).$$

By Theorem 1.17, if (ξ_0, τ_0) is a microlocally elliptic direction over (x_0, t_0), then $u \in H^{2s-(n+1)/2-m}_{ml}(x_0, t_0, \xi_0, \tau_0)$. Hence if $(x_0, t_0) \in \mathbf{R}^{n+1}\setminus C$, it follows that $u \in H^{2s-(n+1)/2-m}_{loc}(x_0, t_0)$. See Figure 1.4.

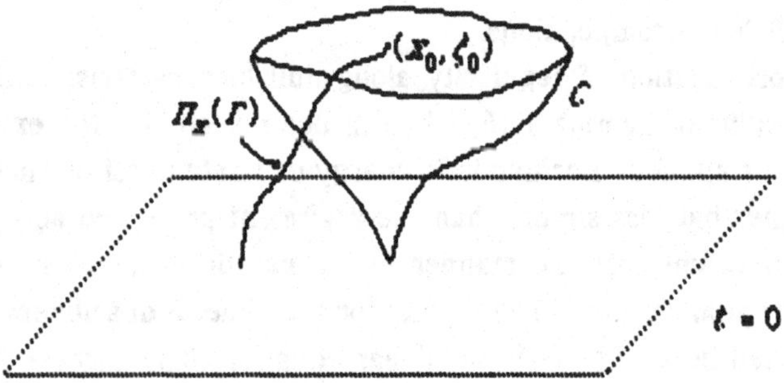

Figure 1.4

In the case of one space dimension, much more precise information is known about the propagation of microlocal regularity, essentially because in that case hyperbolic operators may be factored into products of differential rather than pseudodifferential operators. In particular, for the second order case in one space dimension, Rauch-Reed [58] proved that regularity of order r is propagated along bicharacteristics without any restriction on r. For higher order equations, nonlinear singularities appear in confined locations; their regularity is completely analyzed in Rauch-Reed [59]. In higher dimensions, an index of regularity for solutions to semilinear equations having restricted microlocal singularities is discussed in Chen [26]. It is also important to treat this problem in the case of lower overall regularity; for instance, $s = 1$ corresponds to the classical case of finite energy. A proof of local smoothness for solutions $u \in H^s_{loc}(\mathbf{R}^{n+1}) \cap L^\infty(\mathbf{R}^{n+1})$ for $0 \leq s \leq n/2$ is given in Gerard-Rauch [33].

The central themes that have been described above will appear frequently in the subsequent chapters. Solutions to the linear wave equation will be used to construct solutions to nonlinear problems exhibiting nonlinear singularities. Appropriate microlocal Sobolev spaces will be essential to the measurement of singularities, and their algebra properties will allow for nonlinear analysis. Smoothness in these spaces for solutions to nonlinear equations will usually be established by use of an appropriate exact or approximate commutator argument, coupled with a linear regularity result.

Chapter II. Appearance of Nonlinear Singularities

We have seen that the microlocal regularity of sufficiently smooth solutions to nonlinear strictly hyperbolic equations propagates along null bicharacteristics as in the linear case, as long as that regularity is at most approximately twice the overall smoothness of the solution. That this result is essentially best possible was first demonstrated in Rauch-Reed [58]. They constructed an example of a simple third order semilinear system in one space variable, with operators ∂_t, $\partial_t - \partial_x$, and $\partial_t + \partial_x$, and an explicit solution for which the singular support for $t < 1$ consists of the two characteristic lines $\{x = t\}$ and $\{x = -t\}$, while for $t > 1$ the solution is singular on the additional characteristic $\{t = 0\}$. As is easily deduced from Hörmander's Theorem, in the linear case the only possible singularities would be along the two original lines. Their result is stronger than merely a microlocal statement, since the singular support of the solution, not just its wavefront set, is strictly larger in the nonlinear case than the solution of the linear homogeneous problem with the same initial data. Their result is easily adapted to the case of a single semilinear equation for the third order operator $P = \partial_t(\partial_t - \partial_x)(\partial_t + \partial_x)$. We will begin our consideration of the presence of singularities in nonlinear problems which are absent in the linear case by treating an example for this operator using ideas and techniques which are useful in higher dimensions.

Let E be the forward fundamental solution for P starting at $t = -1$, that is,

$$PEf = f,\ Ef\big|_{t=-1} = 0,\ \partial_t Ef\big|_{t=-1} = 0,\ \partial_t^2 Ef\big|_{t=-1} = 0\,.$$

Then the linear energy inequality (when the hypothesis are made on the Cauchy data on a space-like initial hypersurface, rather than on an open set) implies that E is a continuous map of $H^s_{loc}(\mathbf{R}^{n+1})$ into $H^{s+2}_{loc}(\mathbf{R}^{n+1})$. Let

v be a solution to the linear equation $Pv = 0$. If u is the solution to

$$Pu = f(u),\ u|_{t=-1} = v|_{t=-1},\ \partial_t u|_{t=-1} = \partial_t v|_{t=-1}, \partial_t^2 u|_{t=-1} = \partial_t^2 v|_{t=-1},$$

then we may write

$$u = v + Eu^2 = v + E(v + Eu^2)^2 = v + Ev^2 + 2E(vEu^2) + E(Eu^2)^2.$$

Since E propagates $WF(v^2)$ outwards, it is possible that Ev^2 may have strictly larger singular support than v.

We consider v with very simple singularities across two of the characteristic lines for P: if H is the Heaviside function as before,

$$v = H(x-t)(x-t)^k + H(x+t)(x+t)^k.$$

Then $Pv = 0$, and

$$v^2 = H(x-t)(x-t)^{2k} + H(x+t)(x+t)^{2k} + f,$$
$$f(t,x) = 2H(x-t)H(x+t)(x-t)^k(x+t)^k.$$

The function $f(t,x)$ has strictly larger wavefront over the origin than v. Indeed, from Chapter I,

$$WF(v) = \{(t,x,\tau,\xi): t = x,\ \tau = -\xi\} \cup \{(t,x,\tau,\xi): t = -x,\ \tau = \xi\},$$

while $(\partial_x - \partial_t)^{k+1}(\partial_x + \partial_t)^{k+1}(H(x-t)H(x+t)(x-t)^k(x+t)^k)$ is a constant multiple of δ_0, so $WFv^2 \supset \{(0,0,\tau,\xi): (\tau,\xi) \neq 0\}$. Moreover, $\Box \partial_t Ef = f$, so, as is easily computed, $\partial_t Ef = 1/2\int_R f$, where R is the rectangle as shown in Figure 2.1. A straightforward calculation yields that

$$Ef = c_0 H(x-t)H(x+t)[\varphi(t,x) + |x|^{2k+3}],$$

with $\varphi \in C^\infty(\mathbf{R}^2)$.

If we set $s = k + 1/2 - \varepsilon$ for any small $\varepsilon > 0$, then $v \in H^s_{loc}(\mathbf{R}^2)$. (It will be assumed that $k \geq 1$, so that $s > (n+1)/2$ for $n = 1$.) Then $|x|^{2k+3} \notin H^{2k+3+1/2}{}_{loc}(\mathbf{R}^2) = H^{2s+5/2+2\varepsilon}{}_{loc}(\mathbf{R}^2)$ on any neighborhood of the line $\{x = 0\}$, and hence

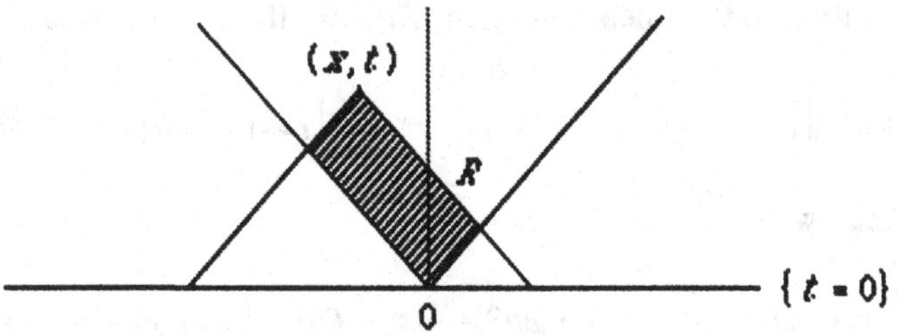

Figure 2.1

$$Ef \notin H^{2s+5/2+2\epsilon}_{loc}((x = 0,\ t > 0)). \tag{2.1}$$

Let $\beta \in C^\infty(\mathbf{R}^2)$ satisfy $\beta = 0,\ t \leq -1/2,\ \beta = 1,\ t \geq 0$. Let u be the solution to

$$Pu = \beta u^2,\ u\big|_{t=-1} = v\big|_{t=-1},\ \partial_t u\big|_{t=-1} = \partial_t v\big|_{t=-1},\quad \partial_t^2 u\big|_{t=-1} = \partial_t^2 v\big|_{t=-1}. \tag{2.2}$$

Then from the calculations above,

$$u = v + E\beta\,[H(x-t)(x-t)^{2k} + H(x+t)(x+t)^{2k}] + E\beta f + 2E\beta(vE\beta u^2) + E\beta(E\beta u^2)^2. \tag{2.3}$$

Sing supp$(v + E\beta[H(x-t)(x-t)^{2k} + H(x+t)(x+t)^{2k}])$ is easily computed to be $\{x = t\} \cup \{x = -t\}$. In order to show that u has larger singular support than the solution of the corresponding linear problem $\Box v = 0$ with the same data at $t = -1$, from (2.3) and (2.1) it is enough to prove that the terms $E\beta(vE\beta u^2)$ and $E\beta(E\beta u^2)^2$ are strictly smoother than $E\beta f$ near $\{x = 0,\ t > 0\}$. Since E is a smoothing operator, it is reasonable to hope that this is the case. Let $\Gamma = \{(t,0,0,\xi_0):\ \xi_0 \neq 0\}$ be the null bicharacteristic corresponding to ∂_t through a point $(t_0, 0)$ with $t_0 > 0$. Then the proof of Theorem 1.11 implies that $u \in H^s_{loc}(\mathbf{R}^2) \cap H^{2s+1}_{ml}(\Gamma)$. (Since the nonlinear function depends only on u and not on Du, microlocal regularity is propagated up to order $r < 2s - (n+1)/2 + 2$ for this third order equation.) By the algebra property, $u^2 \in H^s_{loc}(\mathbf{R}^2) \cap H^{2s-1}_{ml}(\Gamma)$, and thus $(E\beta u^2) \in H^{s+2}_{loc}(\mathbf{R}^2) \cap H^{2s+1}_{ml}(\Gamma)$. Since $v \in H^s_{loc}(\mathbf{R}^2) \cap H^\infty_{ml}(\Gamma)$, it follows from Corollary 1.10 and its extension to functions of differing regularities that

$$(E\beta u^2)^2 \in H^{s+2}{}_{loc}(\mathbf{R}^2) \cap H^{2s+1}{}_{ml}(\Gamma),\ vE\beta u^2 \in H^s{}_{loc}(\mathbf{R}^2) \cap H^{2s+1}{}_{ml}(\Gamma).$$

Therefore, $E\beta(vE\beta u^2) + E\beta(E\beta u^2)^2 \in H^{s+2}{}_{loc}(\mathbf{R}^2) \cap H^{2s+3}{}_{ml}(\Gamma)$, while from (2.1), $E f \notin H^{2s+5/2+2\varepsilon}{}_{ml}(\Gamma)$. Hence from (2.3), $u \notin H^{2s+5/2+2\varepsilon}{}_{ml}(\Gamma)$ as long as $\varepsilon < 1/4$. In particular, the singular support of u contains the forward characteristic $\{x = 0,\ t > 0\}$. See Figure 2.2.

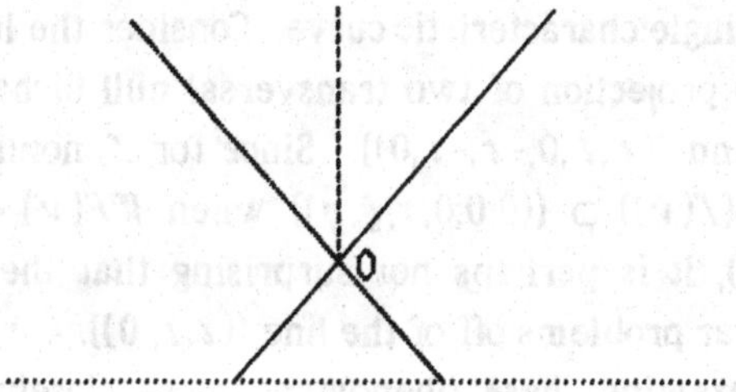

Figure 2.2

Theorem 2.1. *Let* $\beta \in C^\infty(\mathbf{R}^2)$, $\beta = 0$, $t \le -1/2$, $\beta = 1$, $t \ge 0$. *For any* $s > 1$ *there is a choice of Cauchy data* (g_0, g_1, g_2) *such that the solution* $v \in H^s{}_{loc}(\mathbf{R}^2)$ *to*

$$\partial_t(\partial_t - \partial_x)(\partial_t + \partial_x)v = 0,\quad v\big|_{t=-1} = g_0,\ \partial_t v\big|_{t=-1} = g_1,\ \partial_t^2 v\big|_{t=-1} = g_2$$

has sing supp$(v) = \{x = t\} \cup \{x = -t\}$, *while the solution* $u \in H^s{}_{loc}(\mathbf{R}^2)$ *to*

$$\partial_t(\partial_t - \partial_x)(\partial_t + \partial_x)u = \beta u^2,\quad u\big|_{t=-1} = g_0,\ \partial_t u\big|_{t=-1} = g_1,\ \partial_t^2 u\big|_{t=-1} = g_2$$

has sing supp$(u) = \{x = t\} \cup \{x = -t\} \cup \{x = 0,\ t > 0\}$. *Moreover*, $u \notin H^{2s+5/2+\varepsilon}{}_{loc}(\{x = 0,\ t > 0\})$ *for any* $\varepsilon > 0$.

For a second order equation in one space dimension, as discussed earlier, Rauch-Reed [58] prove that no such weak nonlinear singularities can occur. Singularities not present in the linear case arise from the crossing of two or more singularity-bearing characteristics. From each crossing point of singularities for the corresponding linear problem, there are solutions to nonlinear problems with singularities along all of the forward characteristics. (For a second order problem there are no additional characteristics.) Rauch-Reed

[59] prove that these initial and subsequent crossings are the only sources of nonlinear singularities.

On the other hand, in more than one space dimension, nonlinear singularities will in general arise even in second order problems. There are two sources of such singularites. One is the interaction of crossing characteristics (in more than one space dimension there can be more than just the original two characteristic curves issuing from an interaction point). More importantly, a new phenomenon can occur: the "self-spreading" of singularities outward from a single characteristic curve. Consider the line $\{(t,t,0)\} \in \mathbf{R}^3$. It is the (t,x,y) projection of two transversal null bicharacteristics for $\Box$: $\{(t,t,0,\tau,\tau,0)\}$ and $\{(t,t,0,-\tau,-\tau,0)\}$. Since for f nonlinear, it is true in general that $WF(f(v)) \supset \{(0,0,0,\tau,\xi,\eta)\}$ when $WF(v) = \{(t,t,0,\tau,\tau,0)\} \cup \{(t,t,0,-\tau,-\tau,0)\}$, it is perhaps not surprising that the singularities may spread in nonlinear problems off of the line $\{(t,t,0)\}$.

In order to examine these phenomena more closely, we consider the Fourier transforms. In the case of singularities due to crossing, suppose that a solution to the linear wave equation has the simplest possible wave front set associated with the two characteristic curves $\{(t,t,0)\}$ and $\{(t,-t,0)\}$. For instance, take $v \in H^s_{loc}(\mathbf{R}^3)$ with $\Box v = 0$,

$$
\begin{aligned}
WF(v) &= \Gamma_1 \cup \Gamma_2 \\
&= \{(t,t,0,\tau,\xi,\eta) : (\tau,\xi,\eta) \in K_1\} \cup \{(t,0,t,\tau,\xi,\eta) : (\tau,\xi,\eta) \in K_2\}, \\
&\quad K_1 = \{(\tau,\tau,0) : \tau > 0\},\ K_2 = \{(\tau,0,\tau) : \tau > 0\}.
\end{aligned}
$$

Then $v \in H^s_{loc}(\mathbf{R}^3) \cap H^\infty_{ml}((\Gamma_1 \cup \Gamma_2)^{comp})$, while for $f \in C^\infty(\mathbf{R}^3)$ and for $\Gamma = \{(0,0,0,\tau,\xi,\eta) : (\tau,\xi,\eta) \in K_1 + K_2\}$,

$$f(v) \in H^s_{loc}(\mathbf{R}^3) \cap H^\infty_{ml}((\Gamma_1 \cup \Gamma_2 \cup \Gamma)^{comp}).$$

See Figure 2.3a). Since $char(\Box) \cap (\Gamma_1 \cup \Gamma_2 \cup \Gamma) = \Gamma_1 \cup \Gamma_2$, it follows from Hörmander's Theorem that for E the forward fundamental solution for $\Box$ (starting at time $t = -1$, say), $WF(f(v)) \subset \Gamma_1 \cup \Gamma_2 \cup \Gamma$. Then it is not difficult to show that the solution to a nonlinear problem of the form

$$\Box u = f(t,x,u),\ f \equiv 0 \text{ for } t < 0,\ u \equiv v \text{ for } t < 0,$$

satisfies $WF(u) \subset \Gamma_1 \cup \Gamma_2 \cup \Gamma$, and therefore in particular $sing\ supp(u) \subset sing\ supp(v)$. Thus it is necessary to consider the interaction of linear solutions singular on more than a pair of characteristic lines, at least for the restricted case in which the (τ, ξ) projection of the wave front set consists of a single ray above each line.

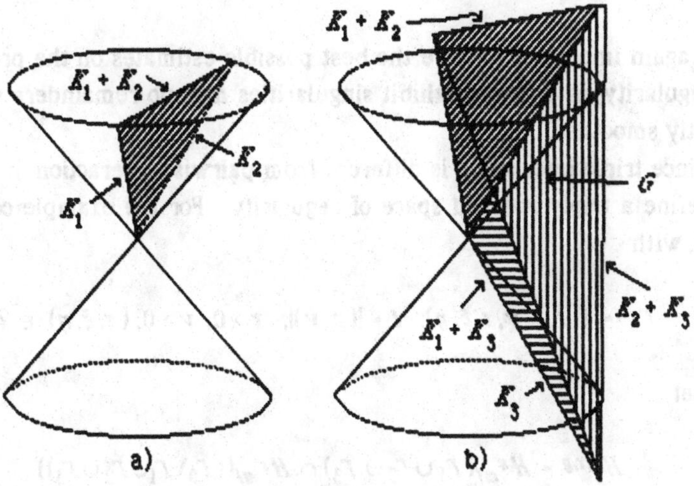

Figure 2.3

For instance, take $v \in H^s_{loc}(\mathbb{R}^3)$ with $\Box v = 0$,

$$WF(v) = \Gamma_1 \cup \Gamma_2 \cup \Gamma_3,$$
$$\Gamma_3 = \{(2^{1/2} t, t, t, \tau, \xi, \eta) \colon (\tau, \xi, \eta) \in K_3\},\ K_3 = \{(-2^{1/2} \tau, \tau, \tau) \colon \tau > 0\}.$$

By decomposing $v^\wedge$ it is easily seen that

$$WF(v^2) \subset \Gamma_1 \cup \Gamma_2 \cup \Gamma_3 \cup B_0,$$
$$B_0 = \{(0,0,0, \tau, \xi, \eta) \colon (\tau, \xi, \eta) \in B = (K_1 + K_2) \cup (K_1 + K_3) \cup (K_2 + K_3)\}.$$

This set contains no new characteristic points, but a third interaction will, in general, spread the singularities to new characteristic directions; for instance, if the Fourier transforms are nonnegative, then

$$WF(v^3) \supset \Gamma_1 \cup \Gamma_2 \cup \Gamma_3 \cup G_4,$$
$$G_4 = \{(0,0,0, \tau, \xi, \eta) \colon (\tau, \xi, \eta) \in G = K_1 + K_2 + K_3\}.$$

See Figure 2.3b). It is then possible for the forward fundamental solution to propagate this larger wavefront set into larger singular support for a solution to a nonlinear equation with corresponding linear solution v, as in the proof of Theorem 2.1. Singularities should spread to

$$\Gamma_4 = \{(t,x,y,\tau,\xi,\eta):\ t = |(x,y)|,\ x > 0,\ y > 0,\ (\tau,\xi,\eta) \in G\}.$$

It is again important to have the best possible estimates on the propagation of regularity, in order to exhibit singularities modulo remainders which are strictly smoother.

Since triple interaction is different from pairwise interaction, it is natural to define a triply indexed space of regularity. For the example considered here, with

$$\Gamma_0 = \{(t,x,y,\tau,\xi,\eta):\ t = |(x,y)|,\ x > 0,\ y > 0,\ (\tau,\xi,\eta) \in B\},$$

we set

$$H^{s,r,g} = H^s{}_{ml}(\Gamma_1 \cup \Gamma_2 \cup \Gamma_3) \cap H^r{}_{ml}(\{\Gamma_0 \backslash \Gamma_1 \cup \Gamma_2 \cup \Gamma_3\})$$
$$\cap\ H^g{}_{ml}(\{\Gamma_4 \backslash \Gamma_0\}) \cap H^\infty{}_{ml}(\{\Gamma_4{}^{comp} \backslash \Gamma_1 \cup \Gamma_2 \cup \Gamma_3\}).$$

From Lemma 1.9 and the results on wavefront sets in proper cones following Definition 1.1, it is easy to show that the space $H^{s,r,g}$ is an algebra for

$$(n+1)/2 < s \leq r < 2s - (n+1)/2,$$
$$r \leq g < \min(r + s - (n+1)/2\ ,\ 3s - (n+1)).$$

Moreover, from Hörmander's Theorem and elliptic regularity (since $\Gamma_0 \backslash \Gamma_1 \cup \Gamma_2 \cup \Gamma_3 \subset (char\,\Box)^{comp}$) it is easily seen that the forward fundamental solution E for $\Box$ maps $H^{s,r,g}$ to $H^{s+1,r+2,g+1}$, at least when acting on functions which are identically zero for $t \leq -1/2$.

Let $\beta \in C^\infty(\mathbf{R}^3)$ satisfy $\beta = 0,\ t \leq -1/2,\ \beta = 1,\ t \geq 0$. Let v be the solution to $\Box v = 0$ given by

$$v\hat{}(t,\xi) = \int e^{i(x\cdot\xi - t|\xi|)} g_1\hat{}(\xi)\, d\xi + \int e^{i(x\cdot\xi - t|\xi|)} g_2\hat{}(\xi)\, d\xi$$
$$+ \int e^{i(x\cdot x + t|\xi|)} g_3\hat{}(\xi)\, d\xi.$$

Here g_1, g_2, and g_3 are defined as in (1.3), (1.4), for σ chosen to put v in $H^s{}_{loc}(\mathbf{R}^3)$, ω_i chosen to make

$$\Pi_{\xi,\eta} WF(g_1) = \{(\xi,0):\ \xi > 0\},\ \Pi_{\xi,\eta} WF(g_2) = \{(0,\eta):\ \eta > 0\},$$
$$\Pi_{\xi,\eta} WF(g_3) = \{(\tau,\tau):\ \tau > 0\},$$

and ρ chosen near zero to make the set on which bounds from below on the Fourier transform exist which are as large as possible compatible with this condition on the wave fronts. Then $WF(v) = \Gamma_1 \cup \Gamma_2 \cup \Gamma_3$. Let u be the solution to

$$(2.4) \qquad \Box u = \beta u^3,\ u\big|_{t=-1} = v\big|_{t=-1},\ \partial_t u\big|_{t=-1} = \partial_t v\big|_{t=-1}.$$

As in the proof of Theorem 2.1, we may write the solution to (2.4) as

$$u = v + E\beta u^3 = v + E\beta v^3 + 3E\beta(v^2 E\beta u^3) + 3E\beta(v(E\beta u^3)^2) + E\beta\,(E\beta u^3)^3.$$

The larger wavefront set $(\Gamma_1 \cup \Gamma_2 \cup \Gamma)$ is present in v^3, by the nonnegativity of the Fourier transform, and is propagated outward by E. It may be explicitly computed that $sing\ supp(E\beta v^3) = \Pi_{t,x,y}(\Gamma_1 \cup \Gamma_2 \cup \Gamma_3 \cup \Gamma_4)$, and the maximal microlocal Sobolev regularity may be computed. The estimates above on the spaces $H^{s,r,s}$ and the argument used in the proof of Theorem 2.1 then apply to show that the remaining terms are smoother remainders. (See Beals [4]). The following result summarizes this example of the spreading of singularities due to crossing. See Figure 2.4.

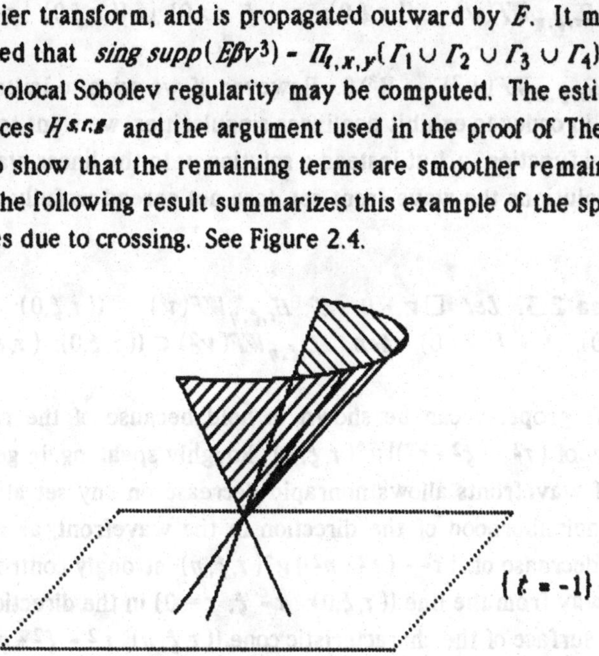

Figure 2.4

Theorem 2.2. *Let $\beta \in C^\infty(\mathbf{R}^3)$, $\beta = 0$, $t \le -1/2$, $\beta = 1$, $t \ge 0$, and let $\varepsilon > 0$ be given. For any $s > 3/2$ there is a choice of Cauchy data (g_0, g_1) such* *that the solution* $v \in H^s_{loc}(\mathbf{R}^3)$ *to* $\Box v = 0$, $v|_{t=-1} = g_0$, $\partial_t v|_{t=-1} = g_1$, *satisfies* sing supp$(v) = \{x = t\} \cup \{y = t\} \cup \{x = y = 2^{-1/2}t\}$, *while the solution* $u \in H^s_{loc}(\mathbf{R}^3)$ *to* $\Box u = \beta u^3$, $u|_{t=-1} = g_0$, $\partial_t u|_{t=-1} = g_1$, *satisfies* sing supp$(u) =$ sing supp$(v) \cup \{t = |(x,y)|, x > 0, y > 0\}$. *Moreover*, $u \notin H^{3s+\epsilon}_{loc}$ *near any point of* $\{t = |(x,y)|, x > 0, y > 0\}$).

The wave front sets considered in the above example are not typical for solutions to the wave equation. For example, singularities along the characteristic line $\{(t,x,0)\colon t = x\}$ will in general correspond to wave front set consisting of the pair of null bicharacteristics

$$\{(t,x,0,\tau,\xi,0)\colon t = x,\ \tau = \xi,\ \tau > 0\} \cup \{(t,x,0,\tau,\xi,0)\colon t = x,\ \tau = \xi,\ \tau < 0\}$$

for $\Box$. As was shown after Definition 1.1, in general if

$$\Pi_{\tau,\xi,\eta}WF(u) = \{(\tau,\xi,0)\colon \tau = \xi,\ \tau > 0\} \cup \{(\tau,\xi,0)\colon \tau = \xi,\ \tau < 0\},$$

then $\Pi_{\tau,\xi,\eta}WF(u^2) = \mathbf{R}^3\backslash 0$. However, if we wish to imitate the argument above in order to exhibit nonlinear singularities, we want to consider not the general function u, but instead a solution v to the linear wave equation. For such solutions the wave front set does not spread as fully as in the general case.

Lemma 2.3. *Let* $\Box v = 0$. *If* $\Pi_{\tau,\xi,\eta}WF(v) = \{(\tau,\xi,0)\colon \tau = \xi,\ \tau > 0\} \cup \{(\tau,\xi,0)\colon \tau = \xi,\ \tau < 0\}$, *then* $\Pi_{\tau,\xi,\eta}WF(v^2) \subset \{(\tau,\xi,0)\colon (\tau,\xi) \neq 0\}$.

This property can be shown to hold because of the rapid decrease at infinity of $[\tau^2 - (\xi^2 + \eta^2)]\, v\hat{}(\tau,\xi,\eta)$. Roughly speaking, in general the definition of wavefronts allows nonrapid decrease on any set almost filling out a conic neighborhood of the direction in the wavefront, as in (1.4). But the rapid decrease of $[\tau^2 - (\xi^2 + \eta^2)]\, v\hat{}(\tau,\xi,\eta)$ strongly controls the growth of $v^{2}\hat{}$ away from the line $\{(\tau,\xi,0)\colon \tau = \xi,\ \tau \neq 0\}$ in the direction perpendicular to the surface of the characteristic cone $\{(\tau,\xi,\eta)\colon \tau^2 = \xi^2 + \eta^2,\ \tau \neq 0\}$. On the other hand, once v has been replaced by v^2, there is no longer any such control over growth away from the tangent plane to the characteristic cone, $\{(\tau,\xi,0)\colon (\tau,\xi) \neq 0\}$, and consequently further interaction will in general yield $\Pi_{\tau,\xi,\eta}WF(v^3) = \mathbf{R}^3\backslash 0$. Thus a triply indexed microlocal Sobolev space is again called for, in order to measure the self-interaction of singularities along a single characteristic. In this case, set

$$\begin{aligned}
\Gamma_1 &= \{(t_0, x_0, y_0, \tau, \xi, 0):\ \tau = \xi,\ \tau \neq 0\},\\
B_\pm &= \{(\tau, \xi, \eta):\ |\tau - \xi| + |\eta| \le \varepsilon|(\tau, \xi, \eta)|,\ \pm\tau > 0\},\\
\Gamma_2 &= \{(t_0, x_0, y_0, \tau, \xi, 0):\ (\tau, \xi) \neq 0\},\\
\Gamma_3 &= \{(t_0, x_0, y_0, \tau, \xi, \eta):\ (\tau, \xi, \eta) \neq 0\}.
\end{aligned}$$

Define

$$\begin{aligned}
(2.5)\qquad H^{s,r,g} &= H^s_{loc}(\mathbf{R}^3) \cap H^r_{ml}(\Gamma_2 \backslash \Gamma_1) \cap H^g_{ml}(\Gamma_3 \backslash \Gamma_2)\\
&\quad \cap \{u:\ \langle \pm|(\xi, \eta)| - \tau\rangle^{r-s} \langle(\tau, \xi, \eta)\rangle^s \chi_{B\pm}(\tau, \xi, \eta)\, u\hat{\ }(\tau, \xi, \eta) \in L^2(\mathbf{R}^3)\}.
\end{aligned}$$

These spaces are more restrictive than the usual microlocal ones, since the growth rate of the Fourier transform is controlled on $B_\pm$ as the characteristic cone is approached. They form algebras for appropriate choices of the indices; for details of this argument, see Beals [5].

Lemma 2.4. *Let* $H^{s,r,g}$ *be defined as in* (2.5). *Then* $H^{s,r,g}$ *is an algebra for* $n/2 < s \le r < 2s - n/2$, $(n+1)/2 < r \le g < \min(r + s - n/2\,, 3s - n)$.

More generally, the (τ, ξ, η) projections of Γ_1, Γ_2, and Γ_3 can be chosen to include more directions, subject to the natural conditions forced by the self-spreading indicated above and the spreading due to the interaction of a pair described previously. Moreover, different choices of (τ, ξ, η) projection may be made over each point (t, x, y). For any number of space dimensions, closed conic subsets of $\mathbf{R}^{n+1}$ are chosen as follows. Let

$$\begin{aligned}
K &= K(t, x) \subset \{|\tau| = |\xi|\},\ K_\pm = K \cap \{\pm\tau = |\xi|\},\\
B &= \text{small conic neighborhood of } \{K_+ + K_+\}^{closure} \cup \{K_- + K_-\}^{closure}\\
&\quad \cup \{(L_+ + L_-)^{closure}:\ L_\pm \subset K_\pm \text{ are rays, } -L_+ \cap L_- = \emptyset\}\\
&\quad \cup \{P_L:\ L \subset K_+ \text{ is a ray with } -L \subset K_-\text{, and } P_L \text{ is the}\\
&\qquad\quad \text{tangent plane to the characteristic cone at } L\},\\
B_\pm &= B \cap \{\pm\tau > 0\}, \text{ and}\\
G &= \{B + B\}^{closure}.
\end{aligned}$$

(If B contains any plane P_L, then $G = \mathbf{R}^{n+1}$.) Set

$$\Gamma_1 = \{(t, x, \tau, \xi):\ (\tau, \xi) \in K(t, x)\},$$

$$\Gamma_2 = \{(t,x,\tau,\xi)\colon\ (\tau,\xi) \in B(t,x)\},$$
$$G_{\pm} = \{(t,x,\tau,\xi)\colon\ (\tau,\xi) \in B_{\pm}(t,x)\},$$
$$\Gamma_3 = \{(t,x,\tau,\xi)\colon\ (\tau,\xi) \in G(t,x)\}, \text{ and}$$
$$\Gamma_4 = \{(t,x,\tau,\xi)\colon\ (\tau,\xi) \in \mathbf{R}^{n+1}\setminus G(t,x)\}.$$

In Figure 2.3 a), K consists of the two rays K_1 and K_2, B is a small neighborhood of the wedge between K_1 and K_2, and $G = B$. In Figure 2.3 b), K is composed of the three rays K_1, K_2 and K_3, B is a small neighborhood of the union of the three wedges between the pairs of rays K_i and K_j, and G is the convex hull of B. In Figure 2.5, K is the union of the two rays $K_{\pm}$ in opposite directions, B is a neighborhood of the tangent plane to the surface of the characteristic cone, and $G = \mathbf{R}^{n+1}$.

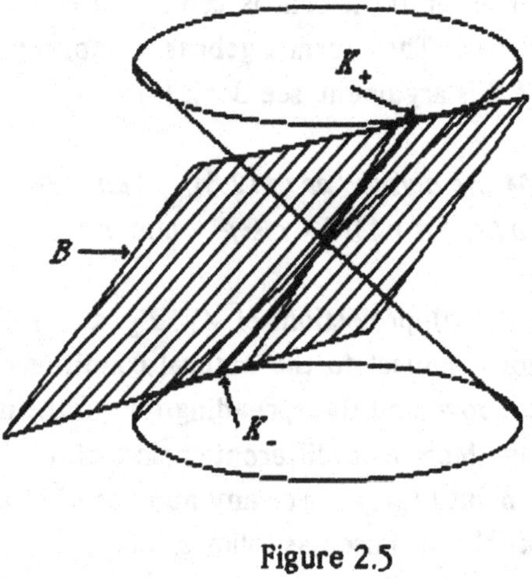

Figure 2.5

Definition 2.5. Let $\Gamma = \{\Gamma_1, \Gamma_2, \Gamma_{\pm}, \Gamma_3, \Gamma_4\}$ be defined as above. Then $u \in H^{s,r,g}(\Gamma)$ if $u \in H^s_{loc}(\mathbf{R}^{n+1}) \cap H^r_{ml}(\Gamma_2\setminus\Gamma_1) \cap H^g_{ml}(\Gamma_3\setminus\Gamma_2) \cap H^\infty_{ml}(\Gamma_4)$ and in addition, $\langle \pm|(\xi,\eta)| - \tau\rangle^{r-s}\langle(\tau,\xi,\eta)\rangle^s \chi_{B\pm}(\tau,\xi,\eta)(\varphi u)^\wedge(\tau,\xi,\eta) \in L^2(\mathbf{R}^3)$ for $\varphi \in C^\infty(\mathbf{R}^{n+1})$ with support sufficiently near (t,x).

Lemma 2.4 is easily seen to remain valid for the spaces $H^{s,r,g}(\Gamma)$. Moreover, there is an analogous result for the case of the product of functions of differing microlocal Sobolev regularity.

In order for the forward fundamental solution of the wave operator to act appropriately on the spaces $H^{s,r,g}(\Gamma)$, the cones $K(t,x)$ must be chosen to allow for the propagation of singularities along forward null bicharacteristics as given by Hörmander's Theorem. A family Γ will be called admissible if the assignment of cones has been made so that $(\tau,\xi) \in K(t_1,x_1)$ whenever $(\tau,\xi) \in K(t_0,x_0)$ and (t_1,x_1,τ,ξ) comes after (t_0,x_0,τ,ξ) on a null bicharacteristic. Then, Hörmander's Theorem and microlocal elliptic regularity easily yield the following result. (Since $\Gamma_2 \backslash \Gamma_1$ avoids the characteristic set for $\Box$, the regularity in the second index is improved by two.)

Theorem 2.6. *Let Γ be an admissible family of conic subsets of $T^*(\mathbf{R}^{n+1})$, and let $H^{s,r,g}(\Gamma)$ be as in Definition 2.5. Let E be the forward fundamental solution for $\Box$ starting at $t = t_0$, and let $\beta \in C^\infty(\mathbf{R}^{n+1})$ have support in $(t \geq t_0)$. If $f \in H^{s,r,g}(\Gamma)$, then $E\beta f \in H^{s+1,r+2,g+1}(\Gamma)$.*

Since the convexity of the regions in (τ,ξ) enclosed by the characteristic half-cones for $\Box$ implies that the wedges $\{K_+ + K_+\}^{closure}$, $\{K_- + K_-\}^{closure}$, and $\{L_+ + L_-\}^{closure}$ and the tangent planes P_L which make up B, only intersect $char(\Box)$ in K itself, the same argument as in Theorem 1.11 allows the conclusion that microlocal regularity up to order $3s - n + 1$ is propagated for H^s solutions of the semilinear wave equation in any number of space dimensions. For higher order equations, the best possible regularity is approximately $2s$, even in one space dimension, as Theorem 2.1 shows. □

Theorem 2.7. *Let Γ be the null bicharacteristic for $\Box$ through the point (t_0,x_0,τ_0,ξ_0). Suppose that $u \in H^s_{loc}(\mathbf{R}^{n+1})$, $s > (n+1)/2$, f is smooth, and $\Box u = f(x,u)$. If $u \in H^r_{ml}(t_0,x_0,\tau_0,\xi_0)$, then $u \in H^r_{ml}(\Gamma)$ for $r < 3s - n + 1$.*

The analogue of this $3s$ result for second order strictly hyperbolic equations holds for the case of variable coefficients, including the general semilinear equation (0.1), and for the quasilinear case (0.2). Stronger restrictions need to be placed on s; the conclusion is again that microlocal H^r regularity is propagated up to order $r < 3s - n + 1 + a$, where a depends on the degree of nonlinearity of the equation. When the coefficients of the principal part $p_2(x,D)$ are variable, the spaces $H^{s,r,g}$ cannot be defined merely in terms of the Fourier transform, as in (2.5). The requirement given there on the size of $\langle \pm|(\xi,\eta)| - \tau\rangle^{r-s}\langle(\tau,\xi,\eta)\rangle^s \chi_{B\pm}(\tau,\xi,\eta)\, u\hat{}(\tau,\xi,\eta)$ is replaced by one on the powers of the operator $p_2(x,D)$ acting on u, essentially

$$[p_2(x,D)]^k u \in H^{s-k}_{loc}(\mathbf{R}^{n+1}) \text{ for } 0 \leq k \leq s - n/2.$$

The key geometric condition that allowed the argument given above to work was the convexity of the regions in (τ, ξ) enclosed by the characteristic half-cones. For a second order equation with variable coefficients the corresponding sets retain this property (Atiyah-Bott-Gårding [3]). Therefore the interaction of a pair of H^s singularities with Fourier transforms essentially supported on $char(p_2)$ will still result in no new singularities in $char(p_2)$. For details in the general case, see Beals [7], Liu [44], Chemin [23].

As in the case of the spreading of singularities due to characteristics crossing, this $3s$ regularity is the best possible result for singularities due to self-spreading. In Beals [5], the proof outlined above for Theorem 2.2 is adapted to this case. A solution v to $\Box v = 0$ is constructed with minimal wavefront set allowing for self-spreading:

$$WF(v) = \{(t, x_1, 0, \tau, \xi_1, 0): t = x_1, \tau = \xi_1, \tau > 0\}$$
$$\cup \{(t, x_1, 0, \tau, \xi_1, 0): t = x_1, \tau = \xi_1, \tau < 0\},$$

and with maximal estimates from below on the size of $v^\wedge$. The nonlinear solution u to

$$\Box u = \beta u^3, \; u|_{t=-1} = v|_{t=-1}, \; \partial_t u|_{t=-1} = \partial_t v|_{t=-1},$$

is written as $u = v + E\beta u^3$, and it can be shown that $E\beta v^3$ has singularities on the entire surface of the forward light cone over the origin, $\{t = |x|\}$. The optimal regularity estimate, Theorem 2.7, is used to show that $E\beta(u^3 - v^3)$ is a strictly smoother remainder. The precise statement of this result on the self-spreading of singularities is as follows.

Theorem 2.8. *Let $n \geq 2$, $s > (n+1)/2$, $\omega_0 \in S^{n-1}$, $0 < t_0 < 1$, $0 < \delta < t - t_0$, and $0 < \varepsilon < 1/2$ be given. Then there exists $\beta \in C^\infty(\mathbf{R})$, with $supp(\beta) \subset \{t: |t - t_0| < \delta\}$ and there is a choice of Cauchy data $\{g_0, g_1\}$ such that the solution $v \in H^s_{loc}(\mathbf{R}^{n+1})$ to $\Box v = 0$, $v|_{t=0} = g_0$, $\partial_t v|_{t=0} = g_1$ has $sing\, supp(v) = \{(t, -t\omega_0): t \in \mathbf{R}\}$, while the solution $u \in H^s_{loc}((-1,1) \times \mathbf{R}^n)$ to $\Box u = \beta u^3$, $u|_{t=-1} = g_0$, $\partial_t u|_{t=-1} = g_1$ has $sing\, supp(u) \supset \{(t, (t - t_0)\omega_0): \delta \leq t - t_0 < 1\}$. Moreover, for any $\chi \in C^\infty(\mathbf{R}^{n+1})$ with support sufficiently near $(t, (t - t_0)\omega_0)$ with $\delta \leq t - t_0 < 1$, $\chi u \notin H^{3s-n+2+\varepsilon}(\mathbf{R}^{n+1})$.*

In particular, *sing supp*(u) contains points in the interior of the solid light cone $\{(t, x): |x| \leq t\}$ over the singular support of the initial data, as in Figure

2.6. In order to show that nonlinear solutions exist with singularities filling the solid light cone, the following condensation of singularities argument applies.

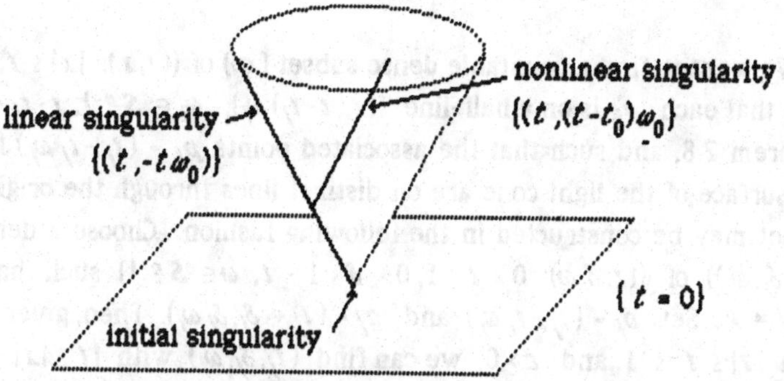

Figure 2.6

First we define an admissible family of cones, as in the case of Theorem 2.6. For $0 \le t \le 1$, set

$$K(x,t) = \emptyset \text{ if } |x| \neq t,$$
$$K(x,t) = \{(\rho, -\rho x/|x|): \rho \neq 0\} \text{ if } 0 < t = |x|, \text{ and}$$
$$K(0,0) = \{(\rho, \rho\omega): \omega \in S^{n-1}, \rho \neq 0\}.$$

Let Γ be the corresponding family of conic subsets of $T^*([0,1] \times \mathbf{R}^n)$ and let $H^{s,r,g}(\Gamma)$ be given by Definition 2.5. Let $s > (n+1)/2$, $0 < \varepsilon < 1/2$, and let $\beta \in C^\infty(\mathbf{R})$ have $supp(\beta) \subset \{0 < t < 1\}$. Suppose that $u \in H^s_{loc}([0,1] \times \mathbf{R}^n)$ satisfies

$$\Box u = \beta u^3, \; u\big|_{t=0} = g_0, \; \partial_t u\big|_{t=0} = g_1, \text{ and } sing\,supp(g_0, g_1) = \{0\}.$$

Let v satisfy $\Box v = 0$, $v\big|_{t=0} = g_0$, $\partial_t v\big|_{t=0} = g_1$, and let E be the forward fundamental solution for $\Box$ starting at $t = 0$. Then

$$u = v + E(\beta u^3) = v + E(\beta v^3) + E(\beta(u^3 - v^3)),$$

and the proof of Theorem 2.7 yields that

$$u - (v + E(\beta v^3)) \in H^{s+1, 2s-n/2+2, 3s-n+5/2-\varepsilon}(\Gamma).$$

In particular,

(2.6) $\chi[u - (v + E(\beta v^3))]\; u \in H^{3s-n+5/2-\epsilon}(\mathbf{R}^{n+1})$ if $\chi \in C^\infty(\mathbf{R}^{n+1})$ has sufficiently small support near a point (t,x) with $|x| < t$.

We want to find a countable dense subset $\{q_j\}$ of $\{(t,x): |x| \le t, 0 \le t \le 1\}$ such that each q_j is on a half-line $\{(t,(t-t_j)\omega_j),\ \omega_j \in S^{n-1},\ t-t_j < 1\}$, as in Theorem 2.8, and such that the associated points $p_j = (t_j, -t_j\,\omega_j)$ located on the surface of the light cone are on distinct lines through the origin. Such a subset may be constructed in the following fashion. Choose a dense subset $\{(t_j, \delta_j, \omega_j)\}$ of $\{(t,\delta,\omega): 0 < t < 1, 0 < \delta < 1 - t,\ \omega \in S^{n-1}\}$ such that $\omega_j \ne \omega_k$ for $j \ne k$. Set $p_j = (t_j, -t_j\,\omega_j)$ and $q_j = (t_j + \delta_j, \delta_j\,\omega_j)$. Then, given $q = (t,x)$ with $|x| \le t \le 1$, and $\epsilon > 0$, we can find $(t_j, \delta_j, \omega_j)$ with $|t - |x| - t_j| < \epsilon/4$, $||x| - \delta_j| < \epsilon/4$, and $|x/|x| - \omega_j| < \epsilon/4$. Hence

$$|q - q_j| \le |t - |x| - t_j| + ||x| - \delta_j| + |(|x| - \delta_j)\,w_j| + |x|\,|x/|x| - w_j| < \epsilon.$$

For each choice of t_j, ω_j, and δ_j as above, let $\beta_j \in C^\infty(\mathbf{R})$ have support in $\{t: |t - t_j| < \delta_j\}$ and let initial data be chosen as in Theorem 2.8, and let the corresponding solution to $\Box v = 0$ be denoted by v_j. Then the solution u_j to $\Box u = \beta_j u^3$ with the same initial data satisfies $\chi_j u \notin H^{3s-n+2+\epsilon}(\mathbf{R}^{n+1})$ for χ_j with support sufficiently near q_j. In particular, since the linear solution v_j is smooth away from the surface of the light cone, it is a consequence of (2.6) that

$$\chi_j E\beta_j(v_j)^3 \notin H^{3s-n+2+\epsilon}(\mathbf{R}^{n+1}). \tag{2.7}$$

We will first demonstrate that there is a choice of initial data with corresponding linear solution v such that $\chi_j E\beta_j v^3 \notin H^{3s-n+2+\epsilon}(\mathbf{R}^{n+1})$ for any j. Next we will show that a single smooth function β can be chosen for which $\chi_j E\beta v^3 \notin H^{3s-n+2+\epsilon}(\mathbf{R}^{n+1})$ for any j. It then follows from (2.6) that the solution to $\Box u = \beta u^3$ satisfies $\chi_j u \notin H^{3s-n+2+\epsilon}(\mathbf{R}^{n+1})$ for any j, and hence *sing supp*$(u) \supset \{(t,x): t \le |x|, 0 \le t \le 1\}$.

For $\mathbf{a} = (a_j) \in l^2(\mathbf{Z}_+)$, set

$$V(\mathbf{a}) = \Sigma(j!)^{-1} a_j v_j.$$

Since each $v_j \in H^{s,\infty,\infty}(\Gamma)$, it follows easily that V maps $l^2(\mathbf{Z}_+)$ continuously into $H^{s,\infty,\infty}(\Gamma)$. Since $V(\mathbf{a})$ satifies $\Box V = 0$, from Lemmas 2.3 and 2.4 we obtain $V^2(\mathbf{a}) \in H^{s,2s-n/2-\varepsilon,\infty}(\Gamma)$ and $V^3(\mathbf{a}) \in H^{s,2s-n/2-\varepsilon,3s-n-\varepsilon}(\Gamma)$. Set $C_{j,m} = \{\mathbf{a} \in l^2(\mathbf{Z}_+)\colon \|\chi_j E\beta_j V^3(\mathbf{a})\|_{H^{3s-n+2+\varepsilon}} \leq m\}$.

Lemma 2.9. *$C_{j,m}$ is closed and nowhere dense in $l^2(\mathbf{Z}_+)$.*

Proof. If $\{\mathbf{a}^l\} \subset l^2(\mathbf{Z}_+)$ is a sequence converging to $\mathbf{a}^0$, set

$$w_l = \chi_j E\beta_j V^3(\mathbf{a}^l),\ w_0 = \chi_j E\beta_j V^3(\mathbf{a}^0).$$

Since $V(\mathbf{a}^l)$ converges to $V(\mathbf{a}^0)$ in $H^{s,\infty,\infty}(\Gamma)$, it follows from Lemmas 2.3 and 2.4 that

$$w_l - w_0 = \chi_j E\beta_j(V(\mathbf{a}^l) - V(\mathbf{a}^0))(V^2(\mathbf{a}^l) + 2V(\mathbf{a}^l)V(\mathbf{a}^0) + V^2(\mathbf{a}^0))$$

converges to 0 in $H^{3s-n+1-\varepsilon}(\mathbf{R}^{n+1})$. If $\{\mathbf{a}^l\} \subset C_{j,m}$, then a subsequence of $\{w_l\}$ converges weakly in $H^{3s-n+2-\varepsilon}(\mathbf{R}^{n+1})$, and therefore has limit w_0. Thus $\mathbf{a}^0 \in C_{j,m}$, whence $C_{j,m}$ is closed.

If $C_{j,m}$ contains an open subset of $l^2(\mathbf{Z}_+)$, it contains a sequence $\mathbf{a}^0$ with only finitely many nonzero entries and for which $a^0{}_j \neq 0$. Since $\chi_j E\beta_j V^3(\mathbf{a}^0) \in H^{3s-n+2-\varepsilon}(\mathbf{R}^{n+1})$, it follows that, for any $e(t,x,\tau,\xi) \in S^0{}_{1,0}$ which has small conic support near $(q_j,1,-\omega_j)$ and is microlocally elliptic at $(q_j,1,-\omega_j)$, $w = e(t,x,D)E\beta_j V^3(\mathbf{a}^0) \in H^{3s-n+2-\varepsilon}(\mathbf{R}^{n+1})$. By Hörmander's Theorem, since $E\beta_j V^3(\mathbf{a}^0)$ is identically zero in $\{t < 0\}$, the smoothness of w is determined by that of $\beta_j V^3(\mathbf{a}^0)$ along the backward null bicharacteristic Γ passing through $(q_j,1,-\omega_j)$. The singular support of $V(\mathbf{a}^0)$, and hence of $V^3(\mathbf{a}^0)$, is contained in $\{(t,x)\colon |x| = |t|\}$. Therefore the (t,x) projection of Γ only intersects *sing supp*$(\beta_j V^3(\mathbf{a}^0))$ at $p_j = (t_j,-t_j\omega_j)$. Let $B(\mathbf{a}^0) = a^0{}_j v_j/j!$, and set $G(\mathbf{a}^0) = V(\mathbf{a}^0) - B(\mathbf{a}^0)$. Then $G(\mathbf{a}^0)$ is smooth on a neighborhood of p_j, since v_k has singular support on the line $\{(t,-t\,w_k)\}$ and $\omega_k \neq \omega_j$ for $k \neq j$. Moreover, $V^3(\mathbf{a}^0) = B^3(\mathbf{a}^0) + R(\mathbf{a}^0)$, with

$$R(\mathbf{a}^0) = \{3B^2(\mathbf{a}^0)G(\mathbf{a}^0) + 3B(\mathbf{a}^0)\,G^2(\mathbf{a}^0) + G^3(\mathbf{a}^0)\}.$$

It is a simple consequence of Lemma 2.3 that $R(\mathbf{a}^0) \in H^\infty{}_{ml}((p_j,1,-w_j))$, and so from above, $e(t,x,D)E\beta_j B^3(\mathbf{a}^0) \in H^{3s-n+2-\varepsilon}(\mathbf{R}^{n+1})$. A contradiction then arises from (2.7), since $B(\mathbf{a}^0) = a^0{}_j v_j/j!$ with $a^0{}_j \neq 0$. Q.E.D.

By Lemma 2.9 and the Baire Category Theorem, the complement of the union of the sets $C_{j,m}$ is dense in $l^2(\mathbf{Z}_+)$. In particular, $a \in l^2(\mathbf{Z}_+)$ of arbitrarily small norm may be chosen such that, with $v = \Sigma(j!)^{-1} a_j v_j$,

$$\begin{gathered} v \in H^s{}_{loc}(\mathbf{R}^{n+1}), \Box v = 0, \; v\big|_{t=0} = g_0, \; \partial_t v\big|_{t=0} = g_1, \\ \textit{sing supp}(g_0, g_1) = \{0\}, \\ \chi_j E\beta_j v^3 \notin H^{3s-n+2+\varepsilon}(\mathbf{R}^{n+1}) \text{ for any } j. \end{gathered} \tag{2.8}$$

For $M = \{\beta \in C^\infty(\mathbf{R})$ with support in $0 \le t \le 1\}$, let A be the continuous map from M to $H^{s+1}([0,1] \times \mathbf{R}^n)$ defined by $A(\beta) = E\beta v^3$, with v fixed as in (2.8). If $\{| \; |_k\}$ is a collection of seminorms on M, define $C_{j,k,m} = \{\beta \in M$: $\|\chi_j A(\beta)\|_{H^{3s-n+2+\varepsilon}} \le m|\beta|_k\}$. It is easily seen that $C_{j,k,m}$ is closed in M. If $\delta > 0$, then by (2.8), $\delta\beta_j \notin C_{j,k,m}$, and therefore $C_{j,k,m}$ is nowhere dense in M. Hence by Baire's Theorem there exists such a β with

$$\chi_j E\beta v^3 \notin H^{3s-n+2+\varepsilon}(\mathbf{R}^{n+1}) \text{ for any } j. \tag{2.9}$$

If v in (2.8) is chosen to have small enough norm, then the solution u to $\Box u = \beta u^3$, $u\big|_{t=0} = g_0$, $\partial_t u\big|_{t=0} = g_1$, will exist up to time $t = 1$, and $u \in H^{s+1}([0,1] \times \mathbf{R}^n)$. Since v has singular support on the surface of the light cone over the origin, it follows from (2.6) and (2.9) that

$$\chi_j u \notin H^{3s-n+2+\varepsilon}(\mathbf{R}^{n+1}) \text{ for any } j,$$

that is, u is singular on the truncated solid light cone $\{(t,x): |x| \le t, 0 \le t \le 1\}$. A similar argument allows the extension of this construction to a solution singular on the entire solid light cone.

Theorem 2.10. *For $n \ge 2$, $s > (n+1)/2$, and $\varepsilon > 0$, there is a choice of Cauchy data (g_0, g_1) with sing supp$(g_0, g_1) = \{0\}$, and there is a function $\beta \in C^\infty(\mathbf{R})$ supported in $\{t \le 0\}$ such that the solution $u \in H^s{}_{loc}(\mathbf{R}^{n+1})$ to $\Box u = \beta u^3$, $u\big|_{t=0} = g_0$, $\partial_t u\big|_{t=0} = g_1$, has sing supp$(u) = \{(t,x): |x| \le t\}$. Moreover, $u \notin H^{3s-n+2+\varepsilon}{}_{loc}$ near any point of $\{(t,x): |x| \le t\}$.*

Finite propagation speed implies that the solid light cone obtained in this example is the largest such set of singularities possible. See Figure 2.7. A further condensation of singularities argument implies that given any closed subset C of $\mathbf{R}^n$ there is a choice of data with singular support C and of smooth function β for which the solution to the problem $\Box u = \beta u^3$ with that data has singular support equal to the union of the regions enclosed by the light cones over the points in C, that is, the singular support of u is the domain of influence of C.

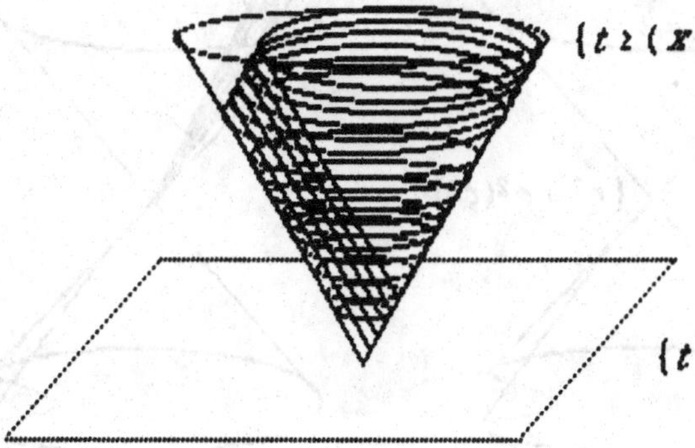

Figure 2.7

Several additional remarks may be made about singularities of types $2s$ and $3s$.

(a) As already has been noted even for the case of one space dimension, if u is an $H^s_{loc}(\mathbf{R}^{n+1})$ solution it is necessary that the strictly hyperbolic equation have order at most two for microlocal regularity of order between approximately $2s$ and $3s$ to be propagated along null bicharacteristics. Moreover, in more than one space dimension, the operator must be differential, not pseudodifferential, because the geometric property that the regions in (τ, ξ) bounded by the two characteristic half-cones be convex is crucial. For example, let $\varphi(\xi) \in C^\infty(\mathbf{R}^n \backslash 0)$ be homogeneous of degree one, strictly positive, and be such that the cone $\{\tau^2 \geq \varphi^2(\xi)\}$ has the nonconvex shape indicated in Figure 2.8a). Microlocal H^r regularity for $H^s_{loc}(\mathbf{R}^{n+1})$ solutions to the strictly hyperbolic pseudodifferential semilinear equation

$$[\partial_t^2 - \varphi^2(\nabla_x)]u = f(t,x,u)$$

will in general not propagate for $r > 2s - (n + 1)/2 + 1$. Indeed, after a single interaction of a pair of crossing singularities, new characteristic directions will be obtained. See Figure 2.8b). The same argument as in the proof of Theorem 2.2 would yield singularities of order $2s - (n+1)/2 + 1$ which are propagated along the projections of the null bicharacteristics passing through these points.

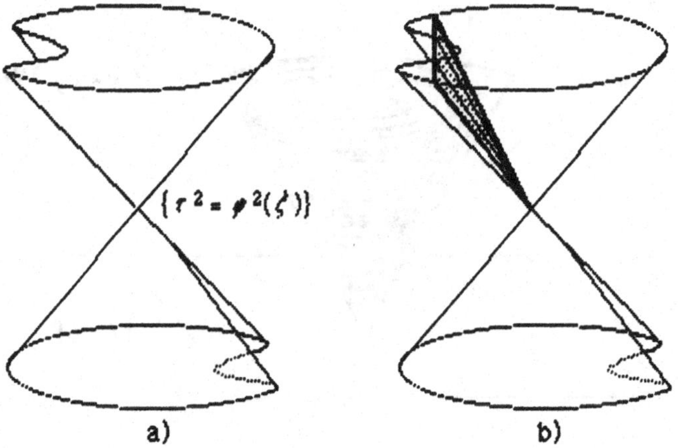

Figure 2.8

(b) The algebra properties of the spaces $H^{s,r,g}(\Gamma)$ given by Definition 2.5 are not invariant under the natural action of Fourier integral operators, that is, under canonical changes of coordinates $(x,\xi) \to (y,\eta)$. For the simpler spaces $H^s_{loc}(\mathbf{R}^{n+1}) \cap H^r_{ml}(\Gamma)$, which do not involve any geometry in the dual variables, the algebra properties are independent of the choice of second coordinate. But in general, the dual geometry is altered under canonical changes, so the spaces $H^{s,r,g}(\Gamma)$ can change. For example, microlocally on the neighborhood B_+ of $\{\tau = |(\xi,\eta)|^2\} \cap \{\tau = \xi\}$, the space $H^{s,r,g}$ defined as in (2.5) but with B_+ alone rather than with $B_\pm$, is an algebra only for $s \leq r \leq n/2$; counterexamples for larger values of r are easily constructed. However, under a standard canonical change of coordinates, B_+ can be transformed into a conic neighboorhood of $\{\tau = 0\} \cap \{\xi = 0\}$, with $\tau - |(\xi,\eta)|$ transformed into τ. The condition defining the space $H^{s,r,g}$ in these coordinates can be shown to define an algebra for all values of $r \geq s$; similar estimates will be discussed in succeeding chapters. The strictly convex characteristic cone is microlocally transformed into a flat hyperplane. Similarly, the characteristic cone of Figure 2.7 may be microlocally changed into the ordinary characteristic cone by a canonical change of coordinates. The corresponding space $H^{s,r,g}$ for the geometry of Figure 2.8 is only an algebra for $g = r$, rather than for $r \leq g < \min(r + s - n/2, 3s - n)$ as in Lemma 2.4.

(c) For higher order strictly hyperbolic equations, the geometry of the characteristic cones determines the optimal microlocal regularity which can be propagated for solutions of nonlinear problems. If $p_m(x, \xi)$ is strictly hyperbolic, then for each fixed x, $\Pi_\xi(\text{char}\,(p_m))\setminus 0$ is the union of m disjoint smooth conic manifolds. Consider three examples. In Figure 2.9a), we treat the case of the product of two wave operators with different speeds of light. Self-interaction of singularities corresponding to a line on the two inner surfaces would in general result in new nonlinear singularities of strength approximately $2s$, since the tangent plane containing the line intersects the characteristic set in new directions, as in Figure 2.9b). On the other hand, self-interaction of a line on the two outer surfaces would in general result in no new propagated nonlinear singularities of strength approximately $2s$, and instead would only give singularities of order approximately $3s$, since the tangent plane containing the line does not intersect the characteristic set in new directions, as in Figure 2.9c). The interactions due to crossings fall into many categories. No propagated nonlinear singularities would result from the interaction of singularities corresponding to a pair of rays both on the upper inner cone, or both on the lower inner cone, or one on the upper outer cone and one on the lower outer cone (in certain cases). But an interaction of one ray on the upper inner cone and one on the lower inner cone would result in new $2s$ singularities. See Figure 2.9d). The interaction corresponding to one ray on the upper inner cone and one on the upper outer cone (or the analogous lower cones) would result in no propagated singularities in some cases, and one of order $2s$ in others, as indicated in Figure 2.9e). Similarly an interaction corresponding to one ray on the upper inner cone and one on the lower outer cone (or the reverse) would result in one new $2s$ singularity in some case, two in others, as in Figure 2.9f). The same holds for the remaining cases of the interaction due to one ray of singularities on each of the upper and lower outer cones.

In Figure 2.10 is drawn the characteristic surface for a more general strictly hyperbolic operator of order four. An analysis similar to the above can be given. One new feature is present: even in the differential rather than pseudodifferential case, we no longer know that all of the sheets of the characteristic set are convex. Therefore the interaction of singularities corresponding to a pair of rays on a single surface can result in the additional propagation of singularities of order $2s$, as indicated, unlike the preceding

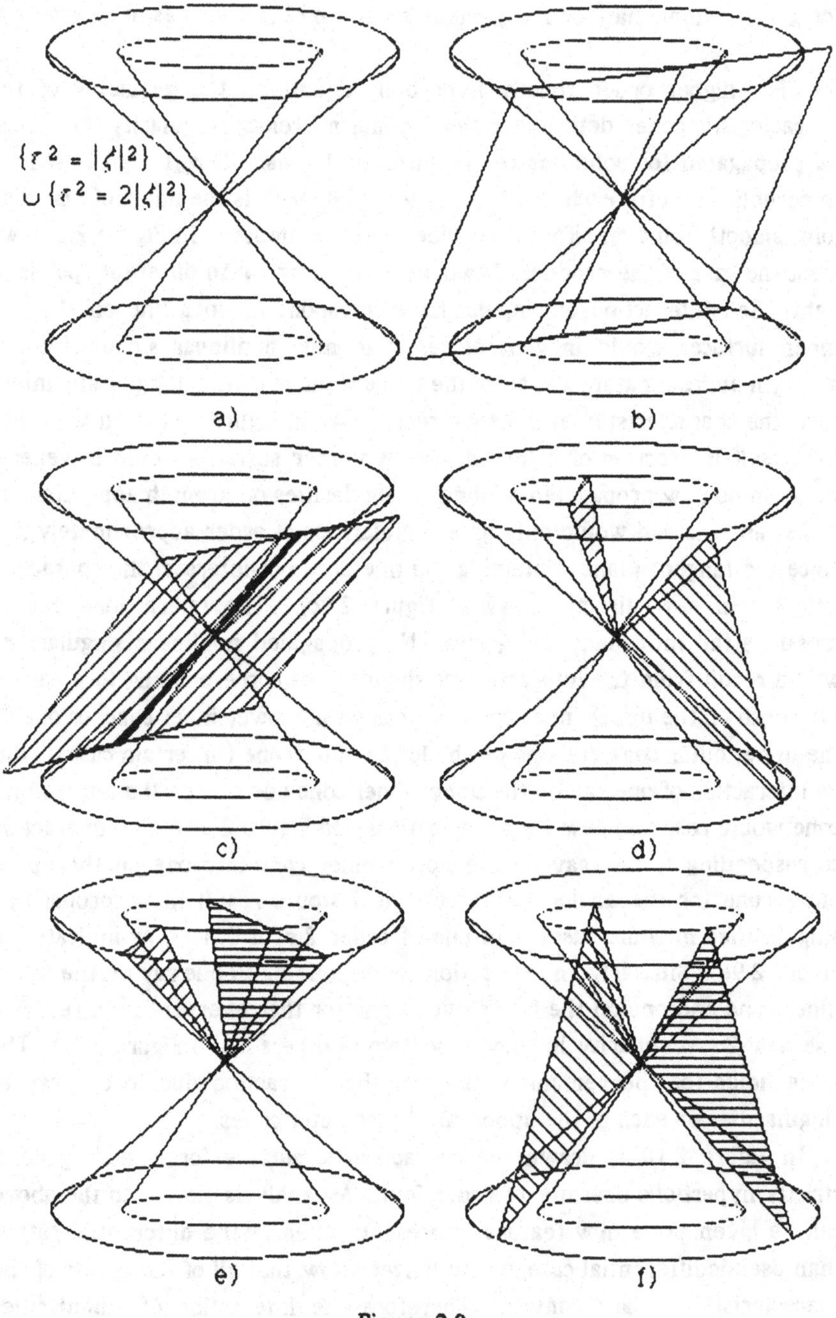

Figure 2.9

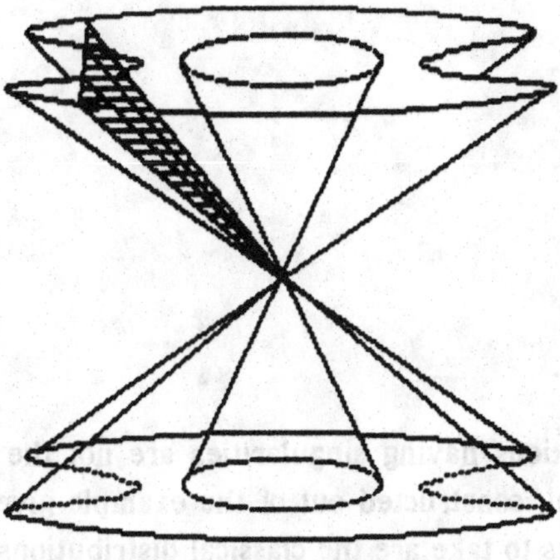

Figure 2.10

case. Examples of such differential operators with non-convex sheets are given in Atiyah-Bott-Gårding [3]. On the other hand, in [3] it is proved that the innermost sheets of the characteristic set for a strictly hyperbolic polynomial are always convex, and therefore such $2s$ singularities will never result from a pair of rays on the uppermost or lowermost sheet.

Chapter III. Conormal Singularities

The simplest functions having singularities are not the ones with prescribed wave front set, constructed out of the example given in (1.3). The natural building blocks to take are the classical distributions with nontrivial singular support; for example, the Dirac distribution $\delta_0(x)$, the Heaviside function $H(x_i)$, or their smoothed versions $|x|^s$, $(x_i)_+{}^r$, $|x_i|^r$, and so on. An appropriate class of functions singular across the hypersurface $x_i = 0$, for example, would be one containing $H(x_i)$ and allowing for multiplication by smooth functions. We are naturally led to the notion of conormal distribution, considered in great generality in Hörmander [37].

Definition 3.1. For $\Sigma \subset \mathbf{R}^n$ a smooth hypersurface, $u \in H^s{}_{loc}(\mathbf{R}^n)$ is said to be conormal with respect to Σ if $M_1 \cdots M_j u \in H^s{}_{loc}(\mathbf{R}^n)$ for all smooth vector fields $M_1, \ldots, M_j$ which are tangent to Σ. If this property holds for all $j \leq k$, u is said to be conormal of degree k with respect to Σ, written $u \in N^{s,k}(\Sigma)$.

An easy consequence of Schauder's Lemma and the chain rule is that the spaces $N^{s,k}(\Sigma)$ are algebras for $s > n/2$. More generally, they are invariant under the action of smooth functions.

Lemma 3.2. *If* $u \in N^{s,k}(\Sigma)$ *for* $s > n/2$, *and* $f(x,v)$ *is a* C^∞ *function of its arguments, then* $f(x,u) \in N^{s,k}(\Sigma)$.

Since the spaces are defined locally, it may be assumed after a local change of coordinates that $\Sigma = \{x_1 = 0\}$. Then $\mathcal{M} = \{x_1\partial x_1, \partial x_2, \ldots, \partial x_n\}$ is a set of generators over $C^\infty(\mathbf{R}^n)$ of all smooth vector fields tangent to Σ. Clearly, $H(x_1) \in N^{s,\infty}(\{x_1 = 0\})$ for all $s < 1/2$, and more generally, $(x_1)_+{}^r \in N^{s,\infty}(\{x_1 = 0\})$ for all $s < r + 1/2$.

It is easily verified that $u \in N^{s,\infty}(\{x_1 = 0\})$ implies that $sing\ supp(u) \subset \{x_1 = 0\}$, since away from that set, $\mathcal{M}$ generates all smooth vector fields. Furthermore, if $N^*\{x_1 = 0\} = \{(0,x_2,\ldots,x_n,\xi_1,0,\ldots,0): \xi_1 \neq 0\}$, the conormal bundle to $\{x_1 = 0\}$, then $u \in N^{s,\infty}(\{x_1 = 0\})$ implies that $WF(u) \subset N^*\{x_1 = 0\}$. Indeed, for $\varphi \in C^\infty_{com}(\mathbf{R}^n)$, $\langle\langle \xi_2,\ldots,\xi_n\rangle\rangle^M(\varphi u)\hat{}(\xi) \in L^2(\mathbf{R}^n)$ for all M, so $\Pi_{\tau,\xi}\, WF(u) \subset \{(\xi_1,0,\ldots,0): \xi_1 \neq 0\}$. In general, then,

$$WF(u) \subset N^*(\Sigma) \text{ for } u \in N^{s,\infty}(\Sigma).$$

As is well known (see for instance Lax [40]), the solution of the "Riemann problem" is fundamental to understanding nonlinear strictly hyperbolic equations and systems in one space variable; this is the initial value problem for data of the form $c_- + c_+H(x)$. For example, if the strictly hyperbolic problem is in the form of a conservation law

$$\partial_t u_i + \partial_x(f_i(u)) = 0,\ i = 1,\ldots,m,$$

then the initial value problem for bounded measurable data of small oscillation and total variation was shown by Glimm to have a solution. (In general the solution will exhibit shocks and rarefaction waves.) Appropriate estimates on solutions of Riemann problems are essential to the proof.

An analogue of this type of problem in higher space dimensions would be the solution of the problem (0.1), (0.2) or (0.3) with inital data conormal with respect to $\{x_1 = 0\}$. Notice that for linear problems, it easily follows from Hörmander's Theorem and the property of wavefronts described above that such an initial value problem will have wavefront set contained in the union of the null bicharacteristics passing over $\{x_1 = 0\}$. In particular, the singular support of such a linear solution to an equation of order m will be contained in the union of the m characteristic hypersurfaces passing through $\{x_1 = 0\}$, as in Figure 3.1.

Since the conormal property is stronger than merely a statement about wavefronts (the space of functions $u \in H^s_{loc}(\mathbf{R}^n)$, $s > n/2$, with $WF(u) \subset N^*\{x_1 = 0\}$ does not form an algebra if $n > 1$), and in particular is well behaved with respect to nonlinear analysis, it might be hoped that the analogous statement holds for solutions to nonlinear problems. Bony [16], [17] established the validity of this property for solutions to semilinear equations. A commutator argument like that described in Chapter I is the key to the proof: regularity for $w = M_1 \cdots M_j u$ is established by examining the

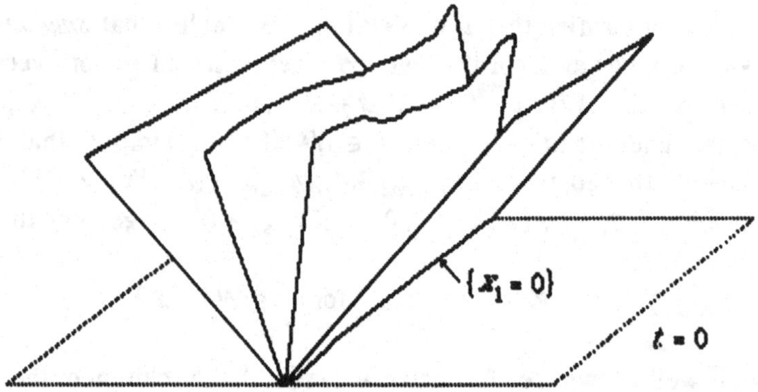

Figure 3.1

action of the operator $p(t,x,D)$ on w. For simplicity we treat the case in which the solution is known to be conormal in the past, rather than on an initial surface. (In the latter case, m characteristic hypersurfaces will come in to play, while in the former only one is present.)

Theorem 3.3. *Let $p_m(t,x,D)$ be a partial differential operator of order m on $\mathbf{R}^{n+1}$ which is strictly hyperbolic with respect to t. Let Σ be a smooth characteristic hypersurface for p_m. Suppose that $u \in H^s_{loc}(\mathbf{R}^{n+1})$ for $s > (n+1)/2 + m - 1$, f is smooth, and $p_m(t,x,D)u = f(t,x,u,\ldots,D^{m-1}u)$. If $u \in N^{s,k}(\Sigma \cap \{t < 0\})$, then $u \in N^{s,k}(\Sigma)$.*

If the nonlinear function depends on fewer derivatives of u, the condition on s may be relaxed. For example, it is sufficient that $u \in H^s_{loc}(\mathbf{R}^{n+1})$, $s > (n+1)/2$, in order for conormal regularity to propagate in solutions of $\Box u = f(t,x,u)$. If, for instance, f is supported in $\{t \geq 0\}$, the unique solution determined by $u = |t - x_1|^r$, $r > n/2$, for $t < 0$ is conormal with respect to $\{t = x_1\}$ as long as it remains in $H^{r+1/2-\epsilon}_{loc}(\mathbf{R}^{n+1})$.

Proof. By a local change of coordinates and finite propagation speed for the operator, we can assume that $\Sigma = \{t = x_1\}$. Then Σ being characteristic for p_m implies that there are operators $q_{m-2}(t,x,\partial_t + \partial x_1, \partial x_2, \ldots, \partial x_n)$, $p_{m-1}(t,x,\partial_t + \partial x_1, \partial x_2, \ldots, \partial x_n)$, and $r_m(t,x,\partial_t + \partial x_1, \partial x_2, \ldots, \partial x_n)$, such that

$$p_m = (t - x_1)(\partial_t - \partial x_1)^2 q_{m-2} + (\partial_t - \partial x_1)p_{m-1} + r_m.$$

The vector fields tangent to Σ are generated by

$$M_0 = (t - x_1)(\partial_t - \partial x_1),\ M_1 = (\partial_t + \partial x_1),\ \text{and}\ M_i = \partial x_i,\ i = 2, \ldots, n.$$

These vector fields commute with each other, and their commutators with p_m are easily seen to satisfy

$$[p_m, M_i] = r_{i,0}(t,x,D)M_0 + \cdots + r_{i,n}(t,x,D)M_n,\ i = 0, \ldots, n,$$

with $r_{i,j}(t,x,D)$ a collection of smooth differential operators of order m-1. If U stands for the vector $(u, M_0 u, \ldots, M_n u)$, then the chain rule and the expressions for the commutators imply that there are smooth vector valued functions g and h and there is a smooth vector valued differential operator $r_{m-1}(t,x,D)$ such that

$$p_m(t,x,D)U = g(t,x,u,\ldots,D^{m-1}u)D^{m-1}U + h(t,x,u,\ldots,D^{m-1}u) + r_{m-1}(t,x,D)U.$$

By assumption, if $k \geq 1$, $U \in H^s{}_{loc}(\{t < 0\})$. Schauder's Lemma and the linear energy inequality then imply that $U \in H^s{}_{loc}(\mathbf{R}^{n+1})$, or equivalently, $u \in N^{s,1}(\Sigma)$. The proof is completed by induction up to order k, with Lemma 3.2 used in place of Schauder's Lemma. Q.E.D.

Now suppose that a solution is known to be conormal in the past with respect to a pair of characteristic hypersurfaces which intersect transversally in the future. Even in one space dimension, it is known that for an equation of order greater than two, the solution to a nonlinear problem will not in general remain conormal with respect to the original pair: the example constructed in Theorem 2.1 was conormal up to time zero. In general, all of the forward characteristic hypersurfaces over the intersection will be the locus of singularities for nonlinear solutions. We first single out the second order case, as in Bony [16]. There, the definition of conormal is easiest, and the appropriate vector fields are no more complicated than the family considered above.

Definition 3.4. For $\Sigma_1, \Sigma_2 \subset \mathbf{R}^n$ a pair of smooth hypersurfaces intersecting transversally, $u \in H^s{}_{loc}(\mathbf{R}^n)$ is said to be conormal with respect to (Σ_1, Σ_2) if $M_1 \cdots M_j u \in H^s{}_{loc}(\mathbf{R}^n)$ for all smooth vector fields $M_1, \ldots, M_j$ simultaneously tangent to both Σ_1 and Σ_2. If this property holds for all $j \le k$, u is said to be conormal of degree k, written $u \in N^{s,k}(\Sigma_1,\Sigma_2)$.

Again, since the regularity is defined in terms of vector fields, the spaces $N^{s,k}(\Sigma_1,\Sigma_2)$ are invariant under the action of smooth functions for $s > n/2$.

Lemma 3.5. *If $u \in N^{s,k}(\Sigma_1,\Sigma_2)$ for $s > n/2$, and $f(x,v)$ is a C^∞ function of its arguments, then $f(x,u) \in N^{s,k}(\Sigma_1,\Sigma_2)$.*

In local coordinates we may take $\Sigma_1 = \{x_1 = 0\}$, $\Sigma_2 = \{x_2 = 0\}$. Then $\mathcal{M} = \{x_1\partial x_1, x_2\partial x_2, \ldots, \partial x_n\}$ is a set of generators over $C^\infty(\mathbf{R}^n)$ of all smooth vector fields simultaneously tangent to Σ_1 and Σ_2. It can be easily verified that $H(x_1)H(x_2) \in N^{s,\infty}(\{x_1 = 0\},\{x_2 = 0\})$ for all $s < 1/2$, and $\delta_0(x_1,x_2) \in N^{s,\infty}(\{x_1 = 0\},\{x_2 = 0\})$ for all $s < -1$. The argument given earlier for a single hypersurface may be adapted to yield *sing supp*$(u) \subset \{x_1 = 0\} \cup \{x_2 = 0\}$ for $u \in N^{s,\infty}(\{x_1 = 0\},\{x_2 = 0\})$, and more precisely,

$$\begin{aligned} WF(u) \subset\ & \{(0,x_2,\ldots,x_n,\xi_1,0,\ldots,0)\colon\ \xi_1 \neq 0\} \\ & \cup \{(x_1,0,\ldots,x_n,0,\xi_2,\ldots,0)\colon\ \xi_2 \neq 0\} \\ & \cup \{(0,0,\ldots,x_n,\xi_1,\xi_2,\ldots,0)\colon\ (\xi_1,\xi_2) \neq 0\}. \end{aligned}$$

In general,

$$WF(u) \subset N^*(\Sigma_1) \cup N^*(\Sigma_2) \cup N^*(\Sigma_1 \cap \Sigma_2) \text{ for } u \in N^{s,\infty}(\Sigma_1,\Sigma_2).$$

The preceding examples show that the wavefront set can be this large.

For a second order equation, no nonlinear singularities appear if the solution is conormal in the past with respect to a pair of smooth characteristic hypersurfaces. See Figure 3.2.

Theorem 3.6. *Let $p_2(t,x,D)$ be a second order partial differential operator on $\mathbf{R}^{n+1}$ which is strictly hyperbolic with respect to t. Let Σ_1 and Σ_2 be smooth characteristic hypersurfaces for p_m which intersect transversally in $\{t \ge 0\}$. Suppose that $u \in H^s_{loc}(\mathbf{R}^{n+1})$, $s > (n+1)/2 + 1$, f is smooth, and $p_2(t,x,D)u = f(t,x,u,Du)$. If $u \in N^{s,k}(\Sigma_1 \cap \{t < 0\}, \Sigma_2 \cap \{t < 0\})$, then $u \in N^{s,k}(\Sigma_1,\Sigma_2)$.*

Proof. Again, by a local change of coordinates and finite propagation speed

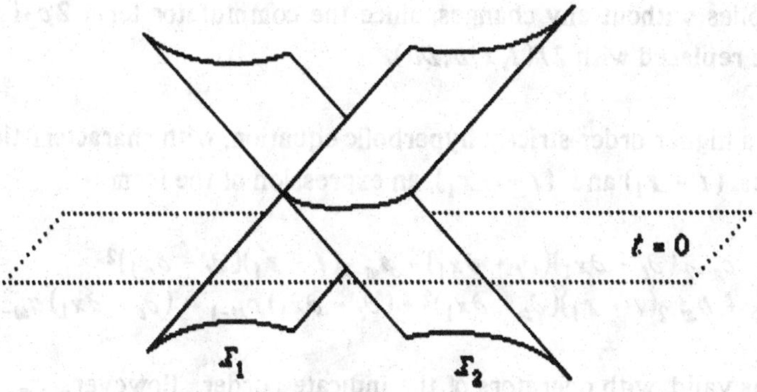

Figure 3.2

for the operator, we can take $\Sigma_1 = (t = x_1)$, $\Sigma_2 = (t = -x_1)$. Then, since both surfaces are characteristic, there are operators $p_1(t, x, \partial x_2, \ldots, \partial x_n)$, $q_1(t, x, \partial x_2, \ldots, \partial x_n)$, and $r_2(t, x, \partial x_2, \ldots, \partial x_n)$ and smooth functions a, b, and c such that

$$p_2 = c(\partial_t - \partial x_1)(\partial_t + \partial x_1) + a(t - x_1)(\partial_t - \partial x_1)^2 + b(t + x_1)(\partial_t + \partial x_1)^2$$
$$+ (\partial_t - \partial x_1) p_1 + (\partial_t - \partial x_1) q_1 + r_2 .$$

By the assumption of strict hyperbolicity, $c \neq 0$, so after division we assume that $c = 1$. The vector fields tangent to both Σ_1 and Σ_2 are generated by

$$M_0 = (t - x_1)(\partial_t - \partial x_1),\ M_1 = (t + x_1)(\partial_t + \partial x_1), \text{ and}$$
$$M_i = \partial x_i,\ i = 2, \ldots, n.$$

These vector fields commute with each other, and as before their commutators with p_2 satisfy

$$[p_2, M_i] = r_{i,0}(t, x, D) M_0 + \cdots + r_{i,n}(t, x, D) M_n ,\ i = 2, \ldots, n,$$

with $r_{i,j}(t, x, D)$ a collection of smooth differential operators of order one. The only difference with the earlier case is in the commutators of p_2 with M_0 and M_1. But, since $[(\partial_t - \partial x_1)(\partial_t + \partial x_1), M_0] = 2(\partial_t - \partial x_1)(\partial_t + \partial x_1)$, it follows that there are smooth differential operators $r_{0,j}(t, x, D)$ with

$$[p_2, M_0] = 2 p_2 + r_{0,0}(t, x, D) M_0 + \cdots + r_{0,n}(t, x, D) M_n ,$$

and a similar expression holds for $[p_2, M_1]$. The rest of the proof of Theorem 3.3 applies without any changes, since the commutator term $2p_2(t,x,D)u$ may be replaced with $2f(t,x,u,Du)$. Q.E.D.

For a higher order strictly hyperbolic equation, with characteristic hypersurfaces $\{t = x_1\}$ and $\{t = -x_1\}$, an expression of the form

$$p_m = c_{m-2}(\partial_t - \partial x_1)(\partial_t + \partial x_1) + a_{m-2}(t - x_1)(\partial_t - \partial x_1)^2 + b_{m-2}(t + x_1)(\partial_t + \partial x_1)^2 + (\partial_t - \partial x_1)p_{m-1} + (\partial_t - \partial x_1)q_{m-1} + r_m$$

remains valid, with operators of the indicated order. However, c_{m-2} is only known to be nonzero microlocally near the union of the conormals to $\{t = x_1\}$ and $\{t = -x_1\}$, so the desired expressions for $[p_m, M_0]$ and $[p_m, M_1]$ in terms of p_m and the M_j are only going to be obtained microlocally (on the set where $c_{m-2}(t, x, \tau, \xi)$ is invertible). Thus microlocal arguments appear in the higher order case. Moreover, the presence of $N^*(\Sigma_1 \cap \Sigma_2)$ in the wavefront of a solution conormal with respect to the pair $\{\Sigma_1, \Sigma_2\}$ implies that, by Hörmander's Theorem, an appropriate conormal space for such solutions is one which allows singularities on all of the surfaces obtained as the projection of the union of all the forward null bicharacteristics over $N^*(\Sigma_1 \cap \Sigma_2)$. If there are m such surfaces $\Sigma_3, \ldots, \Sigma_m$, a definition of conormal with respect to the family $\{\Sigma_1, \ldots, \Sigma_m\}$ is necessary. Vector fields alone are not appropriate to this definition, at least if smooth commutator arguments are to be used, since the vector fields simultaneously tangent to more than two characteristic hypersurfaces intersecting in a codimension two manifold would vanish to high order on that intersection. The corresponding commutators with the strictly hyperbolic operator would not in general be expressible in terms of the operator and the vector fields themselves.

It is natural to consider a microlocal definition of conormal family in the case of this geometry. In general, pseudodifferential operators will not act in a good fashion with respect to the action of nonlinear functions. But Bony [17] established that, for appropriate families of pseudodifferential operators, arguments may be microlocally reduced to the action of vector fields, so that, for instance, the analogue of the chain rule still holds.

Definition 3.7. Let $\Sigma_1, \ldots, \Sigma_m \subset \mathbf{R}^n$ be a family of smooth hypersurfaces intersecting pairwise transversally in the single smooth codimension two submanifold Δ. Then $u \in H^s_{loc}(\mathbf{R}^n)$ is said to be conormal with respect to

$(\Sigma_1, \ldots, \Sigma_m)$ if $M_1 \cdots M_j u \in H^s_{loc}(\mathbf{R}^n)$ for all classical first order pseudodifferential operators $M_1, \ldots, M_j$ with principal symbols vanishing on $N^*\Sigma_1 \cap \ldots \cap N^*\Sigma_m \cap N^*\Delta$. If this property holds for all $j \leq k$, u is said to be conormal of degree k, written $u \in N^{s,k}(\Sigma_1, \ldots, \Sigma_m)$.

For a pair of hypersurfaces this definition is easily seen to agree with the previous one, since then every such pseudodifferential operator M_j may be written as a sum of compositions of operators of order zero with vector fields which are simultaneously tangent to the pair, and operators of order zero preserve $H^s_{loc}(\mathbf{R}^n)$ regularity. The algebra property of such spaces when more than two hypersurfaces are present is proved by a microlocal reduction to such vector fields. The idea is the following: let v_1 and v_2 have microlocal support on sufficiently small cones K_1 and K_2, and suppose we are interested in the microlocal regularity of $M(v_1 v_2)$ on K_3. It suffices to be able to express M microlocally on K_3 as a sum of compositions of operators of order zero with vector fields L_j, with the principal symbols of the terms $\chi_{K_j}(D)L_j$, $j = 1,2$, vanishing on $N^*\Sigma_1 \cap \ldots \cap N^*\Sigma_m \cap N^*\Delta$. Indeed, if that is the case, then the regularity of $\chi_{K_3}(D)M(v_1 v_2)$ is the same as that of $L_j(v_1 v_2) = (L_j v_1) v_2 + v_1(L_j v_2)$. This property, assumed to hold for all triples of sufficiently small cones is called the "three points condition" (Bony [17]). The collection of operators with principal symbols vanishing on $N^*\Sigma_1 \cap \ldots \cap N^*\Sigma_m \cap N^*\Delta$ satisfies the three points condition, since if K_1 and K_2 are sufficiently small, they intersect at most a pair of the conormal spaces $N^*\Sigma_i$ and $N^*\Sigma_j$, and as noted above the operators associated with the pair (Σ_i, Σ_j) are generated by vector fields. By a microlocal decomposition, the following algebra property is then seen to hold.

Lemma 3.8. *Let $\Sigma_1, \ldots, \Sigma_m \subset \mathbf{R}^n$ be a family of smooth hypersurfaces intersecting pairwise transversally in the single smooth codimension two submanifold Δ. If $u \in N^{s,k}(\Sigma_1, \ldots, \Sigma_m)$ for $s > n/2$, and $f(x,v)$ is a C^∞ function of its arguments, then $f(x,u) \in N^{s,k}(\Sigma_1, \ldots, \Sigma_m)$.*

The microlocal commutation property described earlier for a strictly hyperbolic operator of order m with the appropriate collection of pseudodifferential operators associated to the family of m characteristic hypersurfaces, combined with Lemma 3.8, allows the extension of the proof of Theorem 3.3 to the case of a pair of hypersurfaces in the past.

Theorem 3.9 (Bony). *Let $p_m(t,x,D)$ be a partial differential operator of order m on $\mathbf{R}^{n+1}$ which is strictly hyperbolic with respect to t. Let Σ_1 and Σ_2 be smooth characteristic hypersurfaces for p_m which intersect transversally in $\{t \geq 0\}$, and set $\Delta = \Sigma_1 \cap \Sigma_2$. Let $\Sigma_3, \ldots, \Sigma_m$ be the additional forward smooth characteristic hypersurfaces for p_m issuing from Δ. Suppose that $u \in H^s_{loc}(\mathbf{R}^{n+1})$, $s > (n+1)/2 + m - 1$, f is smooth, and $p_m(t,x,D)u = f(t,x,u,\ldots, D^{m-1}u)$. If $u \in N^{s,k}(\Sigma_1 \cap \{t<0\}, \Sigma_2 \cap \{t<0\})$, then $u \in N^{s,k}(\Sigma_1, \ldots, \Sigma_m)$.*

A similar proof handles the initial value problem with data conormal with respect to a single hypersurface, which plays the role of Δ. The conclusion is that the solution is conormal with respect to the family of m characteristic hypersurfaces issuing from Δ, as in Figure 3.1.

More generally, conormal solutions to quasilinear and nonlinear equations (0.2) and (0.3) are treated by Alinhac [1],[2]. In that case, the surfaces are no longer known *a priori* to be smooth in the future; a proof of their regularity is part of the induction procedure. One difference is present from the semilinear case: the surfaces Σ_i are no longer necessarily smooth across Δ, but are smooth outside of Δ (which is itself smooth). Examples exist for which the surfaces are not smooth across Δ even in one space dimension; see Messer [49]. Alinhac uses an extension of Bony's paraproduct of functions to the notion of the paracomposition of functions, in order to change variables to flatten pairs of surfaces. (A further extension to canonical changes of coordinates in the cotangent space and a theory of para-Fourier integral operators is given in Qiu [56].) Similar results may be obtained without changing variables, using instead the techniques of Beals-Reed [14]. We illustrate this procedure in the case of a second order quasilinear equation, with a solution which is assumed to be conormal in the past with respect to a single characteristic hypersurface. For additional details, and for operators of order greater than or equal to two and solutions conormal with respect to a pair of characteristic hypersurfaces in the past, see Kang [38].

Suppose for simplicity of notation that u is a sufficiently smooth solution of the quasilinear strictly hyperbolic equation

$$(3.1)\quad p(t,x,D)u = (\partial_t^2 - \sum_{i,j=1}^{n} a_{i,j}(t,x,u,Du)\partial_{x_i}\partial_{x_j})u = f(t,x,u,Du).$$

If a characteristic hypersurface for p is given locally by $x_1 = \varphi(t,x')$, then φ satisfies the first order nonlinear equation

$$(3.2) \qquad (\varphi_t)^2 = a_{1,1} + \sum_{j=2}^{n}(a_{1,j} + a_{j,1})\varphi_{x_j} + \sum_{i,j=2}^{n} a_{i,j}\varphi_{x_i}\varphi_{x_j}.$$

The coefficients are evaluated at $x_1 = \varphi(t,x')$, and thus the regularity of functions of the form $a(Du(t,\varphi(t,x'),x'))$ needs to be examined.

The vector fields tangent to $x_1 = \varphi(t,x')$ are spanned by

$$(3.3) \qquad \begin{aligned} &M_0 = (x_1 - \varphi(t,x'))\partial x_1,\ M_1 = \partial_t + \varphi_t(t,x')\partial x_1,\\ &M_j = \partial x_j + \varphi_t(t,x')\partial x_1,\ 2 \le j \le n. \end{aligned}$$

The coefficients are not known *a priori* to be regular. The idea of the proof of conormal regularity is to use commutator arguments and (3.1) to establish regularity for u with respect to differentiation by the vector fields given by (3.3) inductively, for coefficients of a given regularity. The improved regularity of u, coupled with (3.2), then yields improved regularity of φ, and hence of the coefficients of the vector fields, allowing the inductive argument to be continued.

We begin by studying the regularity of compositions of the form

$$v(\varphi(x'),x'),$$

with v and φ assumed to have limited Sobolev regularity. A convenient estimate is the following, found in Melrose-Ritter [46]. We treat the case in which the regularity indices s and s' are integers; analogous results hold in the general case.

Lemma 3.10. *Let* $0 \le s' \le s$, *and suppose that* $w \in L^\infty(\mathbf{R}^n) \cap H^s(\mathbf{R}^n)$. *Then for* $|\alpha| = s'$, *it follows that* $D^\alpha w \in L^{2s/s'}(\mathbf{R}^n)$, *and*

$$\|D^\alpha w\|_{L^{2s/s'}} \le C_{s,s',n}(\|w\|_{L^\infty})^{1-s'/s}(\|w\|_{H^s})^{s'/s}.$$

Proof. Let $1 \le j \le s'$, and let D^j stand for any derivative of order j. Then by Hölder's inequality,

$$\begin{aligned} \int (D^j w)^{2s/j}\,dx &= \int (D^j w)^{(2s/j)-1}(DD^{j-1}w)\,dx\\ &= -\int D(D^j w)^{2s/j}(D^{j-1}w)\,dx\\ &= C\int (D^j w)^{(2s/j)-2}(D^{j+1}w)(D^{j-1}w)\,dx\\ &\le C(\int (D^j w)^{2s/j}\,dx)^{1-j/s}\|D^{j+1}w\|_{L^{2s/(j+1)}}\|D^{j-1}w\|_{L^{2s/(j-1)}}, \end{aligned}$$

since $1 - ((j+1)/2s + (j-1)/2s) = 1 - j/s = ((2s/j) - 2)/(2s/j)$. Therefore,

$$\|D^j w\|_{L^{2s/j}} \leq C(\|D^{j+1} w\|_{L^{2s/(j+1)}})^{1/2}(\|D^{j-1} w\|_{L^{2s/(j-1)}})^{1/2},$$

and in particular, $\|Dw\|_{L^{2s}} \leq C(\|w\|_{L^\infty})^{1/2}(\|D^2 w\|_{L^{2s/2}})^{1/2}$. It follows easily by induction on j, using the preceding inequality, that

$$\|D^j w\|_{L^{2s/j}} \leq C(\|w\|_{L^\infty})^{1/(j+1)}(\|D^{j+1} w\|_{L^{2s/(j+1)}})^{j/(j+1)}.$$

Then, it is easily seen by induction on k that

$$\|D^j w\|_{L^{2s/j}} \leq C(\|w\|_{L^\infty})^{k/(j+k)}(\|D^{j+1} w\|_{L^{2s/(j+k)}})^{j/(j+k)}.$$

In particular, with $j = s'$ and $k = s - s'$, the desired estimate follows. Q.E.D.

Corollary 3.11. *Suppose that* $w \in L^\infty{}_{loc}(\mathbf{R}^n) \cap H^s{}_{loc}(\mathbf{R}^n)$ *for* $s \geq 0$. *If* $f \in C^\infty(\mathbf{R})$, *then* $f(w) \in H^s{}_{loc}(\mathbf{R}^n)$.

Proof. If $|\beta| = s$, then by the chain rule and the Leibniz formula, $D^\beta(f(w))$ may be written as a sum of terms of the form $g_\beta(w)D^{\alpha_1}(w) \cdots D^{\alpha_m}(w)$, with g_β smooth and $\alpha_1 + \ldots + \alpha_m = \beta$. Since $g_\beta(w) \in L^\infty{}_{loc}(\mathbf{R}^n)$, and $D^{\alpha_k}(w) \in L^{2s/|\alpha_k|}{}_{loc}(\mathbf{R}^n)$ by Lemma 3.11, Hölder's inequality implies that the use of the chain rule was justified, and since $(|\alpha_1| + \ldots + |\alpha_m|)/2s = 1/2$, $g_\beta(w)D^{\alpha_1}(w) \cdots D^{\alpha_m}(w) \in L^2{}_{loc}(\mathbf{R}^n)$. Q.E.D.

Even if $\varphi(x')$ is smooth, functions of the form $v(\varphi(x'),x')$ will in general have Sobolev regularity of order $1/2$ lower than that of $v(x_1,x')$. When $\varphi(x')$ is nonsmooth, the regularity of $v(\varphi(x'),x')$ will not in general be greater than that of $\varphi(x')$. In order to obtain norm estimates on the regularity of $v(\varphi(x'),x')$ that are linear in the norm of $\varphi(x')$, we will assume that the Sobolev regularity of $v(x_1,x')$ is at least one order greater than that of $\varphi(x')$.

Lemma 3.12. *Let* $v(x_1,x') \in H^{s+1}{}_{loc}(\mathbf{R}^n)$ *for* $s > n/2 + 1$. *Suppose that* $\varphi(x') \in H^{s'}{}_{loc}(\mathbf{R}^{n-1})$, $1 \leq s' \leq s$, *and that* $D\varphi(x') \in L^\infty(\mathbf{R}^{n-1})$. *Then*

$$v(\varphi(x'),x') \in H^{s'}{}_{loc}(\mathbf{R}^{n-1}).$$

If v and φ have compact support, then $\|v(\varphi(x'),x')\|_{H^{s'}} \leq C\|\varphi\|_{H^{s'}}$, with C depending only on n, s, s', the size of the supports, $\|v(x_1,x')\|_{H^{s+1}}$, and $\|D\varphi\|_{L^\infty}$.

Proof. We may assume without loss of generality that v and φ have compact support. Let w represent the vector consisting of all derivatives of $\partial x_1 v$. Then $w \in H^{s-1}(\mathbf{R}^n) \cap L^\infty(\mathbf{R}^n)$ since $s - 1 > n/2$, and therefore

$$D^{\alpha'} w \in L^{2(s-1)/|\alpha'|}(\mathbf{R}^n) \text{ for } |\alpha'| \leq s - 1$$

by Lemma 3.10. Thus $D^\alpha \partial x_1 v \in L^{2(s-1)/(|\alpha|-1)}(\mathbf{R}^n)$ for $1 \leq |\alpha| \leq s$, and since

$$(3.4) \qquad (D^\alpha v)(\varphi(x'),x') = \int_{-\infty}^{\varphi(x')} D^\alpha \partial_{x_1} v(x_1,x')\, dx_1,$$

it follows that $(D^\alpha v)(\varphi(x'),x') \in L^{2(s-1)/(|\alpha|-1)}(\mathbf{R}^{n-1})$ with norm depending only on the size of the supports, $\|v(x_1,x')\|_{H^{s+1}}$, and $\|\varphi\|_{L^\infty}$.

Next, notice that $D\varphi \in H^{s'-1}(\mathbf{R}^{n-1}) \cap L^\infty(\mathbf{R}^{n-1})$, and therefore

$$D^\alpha \varphi \in L^{2(s'-1)/(|\alpha'|-1)}(\mathbf{R}^{n-1}) \text{ for } 1 \leq |\alpha| \leq s'$$

by Lemma 3.10. The chain rule and the Leibniz formula imply that, for $1 \leq |\beta| \leq s'$, $D^\beta(v(\varphi(x'),x'))$ may be written as a sum of terms of the form

$$(D^{\alpha_0} D^m v)(\varphi(x'),x') D^{\alpha_1}\varphi \cdots D^{\alpha_m}\varphi,$$

where D^m stands for a derivative of order m and $\alpha_0 + \ldots + \alpha_m = \beta$. The preceding estimates and Hölder's inequality imply that the use of the chain rule was justified, and

$$(D^{\alpha_0} D^m v)(\varphi(x'),x') D^{\alpha_1}\varphi \cdots D^{\alpha_m}\varphi \in L^2(\mathbf{R}^{n-1})$$

since $(|\alpha_0| + m - 1)/2(s-1) + (|\alpha_1| - 1 + \ldots + |\alpha_m| - 1)/2(s'-1) \leq 1/2$. Moreover, $\|D^\beta(v(\varphi(x'),x'))\|_{L^2}$ is bounded up to an appropriate constant by

$$(\|D\varphi\|_{H^{s'-1}})^{(|\alpha_1| - 1)/2(s' - 1)} \ldots (\|D\varphi\|_{H^{s'-1}})^{(|\alpha_m| - 1)/2(s' - 1)}.$$

Since $|\alpha_1| - 1 + \ldots + |\alpha_m| - 1 \leq s' - 1$, the required estimate holds. Q.E.D.

For solutions of hyperbolic equations, it is natural to consider functions which are continuous in time, with values in appropriate Sobolev spaces in the remaining variables.

Corollary 3.13. *Let* $v(t,x_1,x') \in C(\mathbf{R};H^{s+1}{}_{loc}(\mathbf{R}^n))$ *for* $s > n/2 + 1$. *Suppose that* $\varphi(t,x') \in C(\mathbf{R};H^{s'}{}_{loc}(\mathbf{R}^{n-1})), 0 \leq s' \leq s$, *and that* $D\varphi(t,x') \in C(\mathbf{R};L^\infty(\mathbf{R}^{n-1}))$. *Then* $v(t,\varphi(x'),x') \in C(\mathbf{R};H^{s'}{}_{loc}(\mathbf{R}^{n-1}))$.

Proof. The proof of Lemma 3.12 easily yields that

$$D^\alpha \partial_{x_1} v(t,x_1,x') \in C(\mathbf{R};L^{2(s-1)/(|\alpha|-1)}(\mathbf{R}^n)) \text{ for } 1 \leq |\alpha| \leq s.$$

Therefore, from (3.4) with t as a parameter,

$$(D^\alpha v)(t,\varphi(t,x'),x') - (D^\alpha v)(t,\varphi(t',x'),x') \to 0$$
$$\text{in } L^{2(s-1)/(|\alpha|-1)}(\mathbf{R}^{n-1}) \text{ as } t \to t',$$

and similarly for $(D^\alpha v)(t,\varphi(t',x'),x') - (D^\alpha v)(t',\varphi(t',x'),x')\}$. Therefore

$$(D^\alpha v)(t,\varphi(t,x'),x') \in C(\mathbf{R};\, L^{2(s-1)/(|\alpha|-1)}(\mathbf{R}^{n-1})) \text{ for } 1 \leq |\alpha| \leq s.$$

The rest of the proof of Lemma 3.12 then easily yields continuity of the norm estimates in the parameter t.   Q.E.D.

The defining function for the characteristic hypersurface $\{x_1 - \varphi(t,x')\}$ associated with the quasilinear equation (3.1) satisfies (3.2). We may assume locally that $a_{1,1} \neq 0$, so that φ_t may be expressed as a smooth function of $t,\ x',\ u(t,\varphi(x'),x'),\ Du(t,\varphi(t,x'),x')$, and $D\varphi$ (the x' gradient of φ). Such a function will be denoted by by $f(v(t,\varphi(t,x'),x'),D\varphi(t,x'))$, with v representing the vector (t,x',u,Du).

Lemma 3.14. *Let* $v(t,x_1,x') \in C(\mathbf{R};H^{s+1}{}_{loc}(\mathbf{R}^n))$ *for* $s > n/2 + 1$, *and assume that* $D^2\varphi(t,x') \in L^\infty{}_{loc}(\mathbf{R} \times \mathbf{R}^{n-1})$. *Let* f *be a smooth function of its arguments, and suppose that* $\varphi_t(t,x') - f(v(t,\varphi(x'),x'),D\varphi(t,x'))$. *If* $\varphi(0,x') \in H^s{}_{loc}(\mathbf{R}^{n-1})$, *then* $\varphi(t,x') \in C(\mathbf{R};H^s{}_{loc}(\mathbf{R}^{n-1}))$.

Proof. It can be assumed that the functions in question all have compact support in x'. Let $\varphi^{(s)}$ denote the vector of all x' derivatives of φ up to order s. Under the assumption that φ is smooth, we will establish an *a priori* estimate on the energy $E(t) = \{\int |\varphi^{(s)}(t,x')|^2\, dx'\}^{1/2}$. Standard arguments then allow the smoothness assumption to be dropped.

The chain rule and the Leibniz formula imply that there are smooth functions F and f_α for $|\alpha_1| + \ldots + |\alpha_j| + \ldots + |\alpha_m| \le s$, $j \ge 1$, which are then evaluated at $(v(t,\varphi(x'),x'), D\varphi(t,x'))$, such that

$$\partial_t \varphi^{(s)} = F \cdot D\varphi^{(s)}$$
$$+ \sum f_\alpha D^{\alpha_1}(v(t,\varphi(x'),x')) \cdots D^{\alpha_j}(v(t,\varphi(x'),x')) D^{\alpha_{j+1}}(D\varphi) \cdots D^{\alpha_m}(D\varphi).$$

The energy satisfies $E(t)\,\partial_t E(t) = \int \varphi^{(s)}(t,x')\,\partial_t \varphi^{(s)}(t,x')\,dx'$, and

$$\int \varphi^{(s)} F \cdot D\varphi^{(s)}\,dx' = \int F \cdot D(\varphi^{(s)})^2/2\,dx' = -\int DF \cdot (\varphi^{(s)})^2/2\,dx'.$$

Therefore, from the Cauchy-Schwarz inequality,

$$|E(t)\,\partial_t E(t)| \le C E(t)\,\{\|DF\|_{L^\infty} E(t) + \sum \|f_\alpha D^{\alpha_1}(v(t,\varphi(x'),x'))$$
$$\cdots D^{\alpha_j}(v(t,\varphi(x'),x')) D^{\alpha_{j+1}}(D\varphi) \cdots D^{\alpha_m}(D\varphi)\|_{L^2}\}.$$

Schauder's Lemma implies that $Dv(t,x_1,x') \in L^\infty_{loc}(\mathbf{R} \times \mathbf{R}^n)$, and since $D^2\varphi(t,x') \in L^\infty_{loc}(\mathbf{R} \times \mathbf{R}^{n-1})$, it follows that $D(F(v(t,\varphi(x'),x'), D\varphi(t,x')))$ $\in L^\infty_{loc}(\mathbf{R} \times \mathbf{R}^{n-1})$ and $f_\alpha(v(t,\varphi(x'),x'), D\varphi(t,x')) \in L^\infty_{loc}(\mathbf{R} \times \mathbf{R}^{n-1})$. If

(3.5) $\quad \|D^{\alpha_1}(v(t,\varphi(x'),x')) \cdots D^{\alpha_{j+1}}(D\varphi) \cdots D^{\alpha_m}(Dj)\|_{L^2} \le C(t)E(t),$

then $|\partial_t E(t)| \le C(t)E(t)$. Therefore, by Gronwall's inequality, $E(t)$ is finite for all time (since by assumption it is finite at $t = 0$). Thus the proof is reduced to establishing the validity of (3.5).

By Lemma 3.12 and Corollary 3.13, $v(t,\varphi(x'),x') \in C(\mathbf{R}; H^s(\mathbf{R}^{n-1}))$, with the $H^s(\mathbf{R}^{n-1})$ norm bounded by $C(t)E(t)$. Therefore, $D(v(t,\varphi(x'),x')) \in C(\mathbf{R}; H^{s-1}(\mathbf{R}^{n-1}))$, with similar bounds on the $H^{s-1}(\mathbf{R}^{n-1})$ norm. Moreover, as noted previously, $D(v(t,\varphi(x'),x')) \in L^\infty_{loc}(\mathbf{R} \times \mathbf{R}^{n-1})$, and the $L^\infty(\mathbf{R}^{n-1})$ norm is bounded by $C(t)$. Hence from Lemma 3.10,

$$D^{\alpha_k}(v(t,\varphi(x'),x')) \in C(\mathbf{R}; L^{2(s-1)/(|\alpha_k|-1)}(\mathbf{R}^{n-1})),$$

with the $L^{2(s-1)/(|\alpha_k|-1)}(\mathbf{R}^{n-1})$ norm bounded by $C(t)E(t)^{(|\alpha_k|-1)/(s-1)}$, for $1 \leq |\alpha_k| \leq s$. Similarly,

$$(D\varphi) \in C(\mathbf{R}; H^{s-1}(\mathbf{R}^{n-1})) \cap L^\infty_{loc}(\mathbf{R} \times \mathbf{R}^{n-1}),$$

with the $H^{s-1}(\mathbf{R}^{n-1})$ norm bounded by $E(t)$ and the $L^\infty(\mathbf{R}^{n-1})$ norm bounded by $C(t)$. Thus, again from Lemma 3.10,

$$D^{\alpha_k}(D\varphi) \in C(\mathbf{R}; L^{2(s-1)/(|\alpha_k|)}(\mathbf{R}^{n-1})),$$

with the $L^{2(s-1)/(|\alpha_k|)}(\mathbf{R}^{n-1})$ norm bounded by $C(t)E(t)^{(|\alpha_k|)/(s-1)}$, for $0 \leq |\alpha_k| \leq s-1$. Therefore, by Hölder's inequality,

$$D^{\alpha_1}(v(t,\varphi(x'),x')) \cdots D^{\alpha_j}(v(t,\varphi(x'),x'))D^{\alpha_{j+1}}(D\varphi) \cdots D^{\alpha_m}(D\varphi) \in C(\mathbf{R}; L^2(\mathbf{R}^{n-1})),$$

since $(|\alpha_1| - 1 + \ldots + |\alpha_j| - 1 + |\alpha_{j+1}| + \ldots + |\alpha_m|)/2(s-1) = (s-j)/2(s-1) \leq 1/2$. Moreover, (3.5) holds, because $E(t)^{(s-j)/(s-1)} \leq E(t)$. Q.E.D.

These regularity results and norm estimates are sufficient to establish the analogue of Theorem 3.3 for quasilinear equations. We treat solutions which are in $C(\mathbf{R}; H^s(\mathbf{R}^n)) \cap C^1(\mathbf{R}; H^{s-1}(\mathbf{R}^n))$, and define the conormal spaces accordingly.

Theorem 3.15. *Let* $p(t,x,D)$ *be the second order quasilinear partial differential operator given in* (3.1). *Let* Σ *be a characteristic hypersurface for* p, *defined by a function with nonvanishing gradient. Suppose that* $u \in C(\mathbf{R}; H^s(\mathbf{R}^n)) \cap C^1(\mathbf{R}; H^{s-1}(\mathbf{R}^n))$, $s > n/2 + 3$, f *is smooth, and* $p(t,x,D)u = f(t,x,u,Du)$. *If* Σ *is smooth in* $\{t < 0\}$ *and* $u \in N^{s,k}(\Sigma \cap \{t < 0\})$, *then* Σ *is smooth for all time, and* $u \in N^{s,k}(\Sigma)$.

Proof. By finite propagation speed and an analysis local in time, it may be assumed that u has compact support in x for each t, and that Σ is given by $\{x_1 = \varphi(t,x')\}$. Then φ satisfies (3.2), which is a nonlinear equation of first order with $C^2(\mathbf{R}^n)$ coefficients, by Schauder's lemma and (3.1), because $Du \in C(\mathbf{R}; H^{s-1}(\mathbf{R}^n)) \cap C^1(\mathbf{R}; H^{s-2}(\mathbf{R}^n))$ and $s - 1 > n/2 + 2$. Therefore, $\varphi \in C^2(\mathbf{R} \times \mathbf{R}^{n-1})$. Let $v = (u, Du)$, and set $s_0 = s - 2$. Then, from (3.2), the hypotheses of Lemma 3.14 are satisfied for s_0, and therefore

$$(3.6)\qquad \varphi(t,x') \in C(\mathbf{R};H^{s_0}(\mathbf{R}^{n-1})).$$

The commuting vector fields M_i given by (3.3) have continuous coefficients, and span the collection of all vector fields which are tangent to Σ and have continuous coefficients. From the proof of Theorem 3.3, there are first order operators $a_i(D)$ and $b(D)$ such that $p(t,x,D) = \sum a_i(D)M_i + b(D)$. The coefficients of the terms involving $M_1,\ldots,M_n$ are algebraic expressions involving the coefficients of $p(t,x,D)$ and those appearing in the M_i, and hence are smooth functions of u, Du, φ, and $D\varphi$. The term involving M_0 occurs because the coefficient $a_{1,1}$ of ξ_1^2 in $p(t,x,\xi)$ vanishes at $x_1 = \varphi(t,x')$ by the assumption that this hypersurface is characteristic. Therefore, $a_{1,1}$ may be written as the product of $(x_1 - \varphi(t,x'))$ with an integral of the derivative of the coefficient of ξ_1^2, thus involving u, Du, and D^2u. Consequently, the coefficients of the operators $a_i(D)$ are smooth functions and integrals of smooth functions of u, Du, D^2u, φ, and $D\varphi$, which will written symbolically as $a_i(D^2u,D\varphi,D)$. Similarly, $b(D) = b(D^2u,D^2\varphi,D)$. Because the vector fields commute with each other, their commutators with the operator satisfy

$$[\textstyle\sum a_i(D)M_i,\ M_j] = \sum b_{i,j}(D)M_i,$$

with $b_{i,j}(D) = b_{i,j}(D^3u,D^2\varphi,D)$.

Since the regularity of the coefficients of the vector fields M_i is *a priori* lower than than that of u, it is necessary to differentiate the equation (3.2) in order to use the commutator argument on an appropriate derivative of u. Let U stand for the vector of all derivatives of u up to third order. From (3.1) and the assumptions on u, $U \in C(\mathbf{R};H^{s-3}(\mathbf{R}^n)) \cap C^1(\mathbf{R};H^{s-4}(\mathbf{R}^n))$. Then $p(t,x,D)U = g(U)$ for a smooth function g. (The right hand side is meaningful even though $s-4 < n/2$, because g is linear in the highest order components of U, and the product of a function in $H^{s-3}(\mathbf{R}^n)$ with a function in $H^{s-4}(\mathbf{R}^n)$ is in $H^{s-4}(\mathbf{R}^n)$.) Let M stands for vector operator $(1,M_0,\ldots,M_n)$. The expressions for $p(t,x,D)$ and the commutators given above imply that there are smooth functions G and h such that

$$(3.7)\qquad p(t,x,D)MU = G(U)MU + b_1(D)MU,$$

where $b_1(D)$ is a first order operator with coefficients depending smoothly on U and $D^2\varphi$.

The coefficients of p are in $C(\mathbf{R};H^{s-1}(\mathbf{R}^n)) \cap C^1(\mathbf{R};H^{s-2}(\mathbf{R}^n))$, while by

(3.6) and (3.2) the coefficients of $b_1(D^3 u, D^2 \varphi, D)$ are in $C(\mathbf{R};H^{s-3}(\mathbf{R}^n))$. It then follows from (3.7) and a straightforward modification of the usual proof of the linear energy inequality that

$$(3.8) \qquad MU \in C(\mathbf{R};H^{s-3}(\mathbf{R}^n)) \cap C^1(\mathbf{R};H^{s-4}(\mathbf{R}^n)).$$

In order to use the improved regularity of U, it is convenient to differentiate (3.2) twice. It follows that there are smooth functions f_j such that

$$(D^2\varphi)_t = f_0(Du(t,\varphi,x'),D(D^2\varphi)) + f_1(U(t,\varphi,x'),(D^2\varphi)).$$

Since $(M_1 v)(t,\varphi,x') = \partial_t(v(t,\varphi,x'))$ and $(M_j v)(t,\varphi,x') = \partial x_j(v(t,\varphi,x'))$ for $2 \le j \le n$, it follows that there are smooth functions F_j with

$$(3.9) \qquad (D^3\varphi)_t = F_0(Du(t,\varphi,x'),D(D^3\varphi)) + F_1(MU(t,\varphi,x'),(D^3\varphi)).$$

Let $s_1 = s - 4 = s_0 - 2$. Since $MU \in C(\mathbf{R};H^{s-3}(\mathbf{R}^n))$, it follows from Corollary 3.13 that $MU(t,\varphi,x') \in C(\mathbf{R};H^{s_1}(\mathbf{R}^{n-1}))$. The energy inequality used in the proof of Lemma 3.14 then implies that $D^3\varphi \in C(\mathbf{R};H^{s_1}(\mathbf{R}^{n-1}))$, that is, $\varphi \in C(\mathbf{R};H^{s_0+1}(\mathbf{R}^{n-1}))$.

Suppose inductively that (3.6) is improved to $\varphi \in C(\mathbf{R};H^{s_0+j}(\mathbf{R}^{n-1}))$ and (3.8) is improved to $M^{j-1}U \in C(\mathbf{R};H^{s-3}(\mathbf{R}^n)) \cap C^1(\mathbf{R};H^{s-4}(\mathbf{R}^n))$. From the analogue of (3.9) for $D^{j+1}\varphi$, it follows that

$$\varphi \in C^1(\mathbf{R};H^{s_0+j-1}(\mathbf{R}^{n-1})) \cap \ldots \cap C^j(\mathbf{R};H^{s_0}(\mathbf{R}^{n-1})).$$

Then, an equation similar to (3.7) holds for M^jU, with coefficients depending smoothly on $M^{j-1}U$ and $D^{j+2}\varphi$. Consequently, (3.8) may be improved to $M^jU \in C(\mathbf{R};H^{s-3}(\mathbf{R}^n)) \cap C^1(\mathbf{R};H^{s-4}(\mathbf{R}^n))$. Since an equation of the form (3.9) holds for $D^{j+2}\varphi$, with MU replaced by M^jU, it follows that (3.6) is then improved to $\varphi \in C(\mathbf{R};H^{s_0+j+1}(\mathbf{R}^{n-1}))$. The induction step is complete, and the regularity of the characteristic hypersurface Σ and the conormal regularity of u are established. Q.E.D.

Conormal functions of lower order $H^s(\mathbf{R}^{n+1})$ regularity have also been considered. Ritter [64] treats solutions to semilinear equations of the form $p_2(t,x,D)u = f(t,x,u)$ which are conormal with respect to a single smooth characteristic hypersurface and are contained in $H^s_{loc}(\mathbf{R}^{n+1})$ for any $s >$

1/2. Métivier [50] studies even lower regularity, including discontinuous solutions.

Even for a linear problem, characteristic hypersurfaces may themselves have singularities. For example, the wave operator $\square$ has among its characteristic hypersurfaces the surface of the light cone (with a singularity at its vertex), the cusp, and the swallowtail. The study of solutions conormal in an appropriate sense with respect to such a surface or family of surfaces is naturally more complicated. For solutions to nonlinear problems conormal in the past with respect to the surface of the light cone, or the corresponding initial value problem for data conormal with respect to a single point, the semilinear equation for $\square$ is treated in Beals [6] and [8].

Definition 3.16. For $\Sigma = \{(t,x): |t| = |x|\} \subset \mathbf{R}^{n+1}$, $u \in H^s_{loc}(\mathbf{R}^{n+1})$ is said to be conormal with respect to Σ if $M_1 \cdots M_j u \in H^s_{loc}(\mathbf{R}^{n+1})$ for all smooth vector fields $M_1, \ldots, M_j$ which are tangent to $\Sigma\backslash\{0\}$. If this property holds for all $j \leq k$, u is said to be conormal of degree k with respect to Σ, written $u \in N^{s,k}(\Sigma)$.

Away from the origin, this definition agrees with that given in the case of a single nonsingular hypersurface. A set of generators over $C^\infty(\mathbf{R}^n)$ of all such vector fields is given by

$$(3.10)\quad \mathcal{M} = \{t\partial_t + x_1\partial x_1 + \ldots + x_n\partial x_n,\ t\partial x_j + x_j\partial_t,\ x_i\partial x_j - x_j\partial x_i\}$$

These vector fields are clearly tangent to Σ. In order to verify that they yield all such vector fields, we use induction on n. In one space dimension, clearly $\{t\partial_t + x_1\partial x_1, t\partial x_1 + x_1\partial_t\}$ is equivalent to the known set of generators $\{(t + x_1)(\partial_t + \partial x_1), (t - x_1)(\partial_t - \partial x_1)\}$. Next, notice that if $c(t,x)$ is a smooth function vanishing on $\{t^2 = |x|^2\}$, there is a smooth function $d(t,x)$ such that $c(t,x) = d(t,x)(t^2 - |x|^2)$. Indeed, by the Malgrange Preparation Theorem, there are smooth functions $c_i(x)$ and $d(t,x)$ such that

$$c(t,x) = d(t,x)(t^2 - |x|^2) + c_1(x)t + c_0(x),$$

with $c_1(x)t + c_0(x)$ vanishing on $\{t^2 = |x|^2\}$, whence $\pm|x|c_1(x) = c_0(x)$, and $c_1(x) = c_0(x) = 0$. If $M = a_0\partial_t + a_1\partial x_1 + \ldots + a_n\partial x_n$ is tangent to Σ, then

$$M(t^2 - |x|^2) = a_0 t + a_1 x_1 + \ldots + a_n x_n = d(t,x)(t^2 - |x|^2).$$

It follows by setting $x_n = 0$ that $M_0 = a_0(t,x',0)\partial_t + \ldots + a_{n-1}(t,x',0)\partial x_{n-1}$ is tangent to $\{t^2 - |x'|^2\}$. By the inductive hypothesis, M_0 may be expressed as a smooth combination of ∂x_n and the vector fields in $\mathcal{M}$. Moreover, since $M = M_0 + a_n(t,x',0)\partial x_n + x_n M_1$, and $\{x_n\partial_t + t\partial x_n, x_n\partial x_i - x_i\partial x_n\} \subset \mathcal{M}$, it follows that $M = b\partial x_n$, modulo elements of $\mathcal{M}$. Therefore $b\partial x_n(t^2 - |x|^2) = -x_n b$ vanishes on $\{t^2 - |x|^2\}$, so that $b = d(t,x)(t^2 - |x|^2)$. Finally, we have

$$(t^2 - |x|^2)\partial x_n = -x_n(t\partial_t + x_1\partial x_1 + \ldots + x_n\partial x_n) + t(x_n\partial_t + t\partial x_n) + \ldots$$
$$+ x_{n-1}(x_n\partial x_{n-1} - x_{n-1}\partial x_n).$$

It is natural to consider whether a solution to $\Box u = f(t,x,u)$ which is conormal with respect to the surface of the light cone in the past is conormal with respect to the surface of the light cone after the singular point. Similarly, functions are said to be conormal with respect to the origin if they do not lose regularity when differentiated by a vector field with coefficients vanishing at the origin. If u has data at time $t = 0$ which are conormal with respect to the origin, it is natural to ask whether the resulting solution is again conormal with respect to the surface of the light cone. In particular, this property would imply that no nonlinear singularities are present on the interior of the light cone, unlike the example constructed in Theorem 2.10. It turns out that weaker hypotheses than conormality allow this conclusion. Solutions that are "radially smooth" (that is, they do not lose regularity when acted upon by the radial vector field $t\partial_t + x_1\partial x_1 + \ldots + x_n\partial x_n$) are smooth on the interior of the light cone. We establish this weaker property before dealing with full conormal regularity. The corresponding result for the initial value problem with radially smooth data is proved in a similar fashion.

Theorem 3.17. *Let* $M = t\partial_t + x_1\partial x_1 + \ldots + x_n\partial x_n$, *and suppose that* $u \in H^s_{loc}(\mathbf{R}^{n+1})$, $s > (n+1)/2$, *satisfies* $M^j(u) \in H^s_{loc}(\{t < 0\})$ *for all* j. *If* $\Box u = f(t,x,u)$, *then* $u \in C^\infty(\{(t,x): |x| < |t|\})$.

Proof. The commutator satisfies $[\Box, M] = 2\Box$, so the usual argument easily implies that $M^j(u) \in H^s_{loc}(\mathbf{R}^{n+1})$. In particular,

$$WF(u) \subset \{(t,x,\tau,\xi): t\tau + x\cdot\xi = 0\}. \tag{3.11}$$

If $|x_0| < |t_0|$ and $(t_0,x_0,\tau_0,\xi_0) \in WF(u)$, then $|\xi_0| > |\tau_0|$, so $\Box$ is microlocally elliptic at $(t_0, x_0, \tau_0, \xi_0)$. Thus if there is a neighborhood U of (t_0, x_0) such that $u \in H^r_{loc}(U)$ for $r > s$, then $f(t,x,u) \in H^r_{loc}(U)$, and therefore $u \in H^{r+2}_{ml}(t_0, x_0, \tau_0, \xi_0)$. From (3.11), $u \in H^{r+2}_{loc}(U)$ if U is sufficiently small, so the proof is completed by induction. Q.E.D.

All of the vector fields given by (3.10) except for the radial one commute with $\square$. Therefore (3.10) and the usual commutator argument imply that full conormal regularity with respect to the surface of the light cone persists when it is present in the past.

Theorem 3.18. *Let* $\Sigma = \{(t,x): |t| = |x|\}$. *Suppose that* $u \in H^s_{loc}(\mathbf{R}^{n+1})$, f *is smooth*, $s > (n+1)/2$, *and* $\square u = f(t,x,u)$. *If* $u \in N^{s,k}(\Sigma \cap \{t < t_0\})$, *then* $u \in N^{s,k}(\Sigma)$.

This result also holds for the semilinear equation $\square u = f(t,x,Du)$ if $s > (n+1)/2 + 1$. For the general smooth second order strictly hyperbolic operator $p_2(t,x,D)$, normal coordinates may be chosen locally near a point so that the characteristic cone over that point is the usual one. The commutators of the operator with the vector fields given by (3.10) may then be written as

$$[p_2(t,x,D), M] = a\,p_2(t,x,D) + \sum b_i(t,x,D)\,M_i$$

as in the proof of Theorem 3.6, so that the commutator argument again works locally. Away from the vertex, as long as the characteristic cone remains nonsingular, Theorem 3.3 applies. Therefore the results remain true for the solution of the general second order semilinear strictly hyperbolic equation, as long as the characteristic cone does not develop singularities away from its vertex.

The analogous theorems hold for the initial value problem, if the data are assumed to be conormal with respect to the origin. The restrictions to $\{t = 0\}$ of the vector fields $t\partial_t + x_1\partial x_1 + \ldots + x_n\partial x_n$ and $x_i\partial x_j - x_j\partial x_i$ all vanish at the origin. Multiplication by x_i improves the regularity of a function conormal with repect to the origin by one, so that the restriction to $\{t = 0\}$ of $t\partial x_i + x_i\partial_t$ is also well behaved. The commutator argument for the semilinear problem may therefore be adapted in a straightforward manner. The case of a fully nonlinear strictly hyperbolic problem with initial data conormal with respect to a point is treated in Chemin [24]. As in Theorem 3.15, the difficult part of the proof is to establish that the surface in question is smooth (in this case, away from the vertex).

An analysis of the cusp surface has been given by Ritter, and in a more general setting by Melrose [45]. We consider only a model case, with a weaker notion of conormal regularity than that treated in [45]. Let Σ be the standard cusp $\{(x,y,z): y^3 = x^2\}$. The operator

$$p_2(D) = \partial_x\partial_z - \partial_y^2 + (9y/4)\partial_x^2$$

is strictly hyperbolic with respect to the time variable $t = x + z$ on a neighborhood of the origin, since after a linear change of coordinates it is the d'Alembertian at the origin. See Figure 3.3.

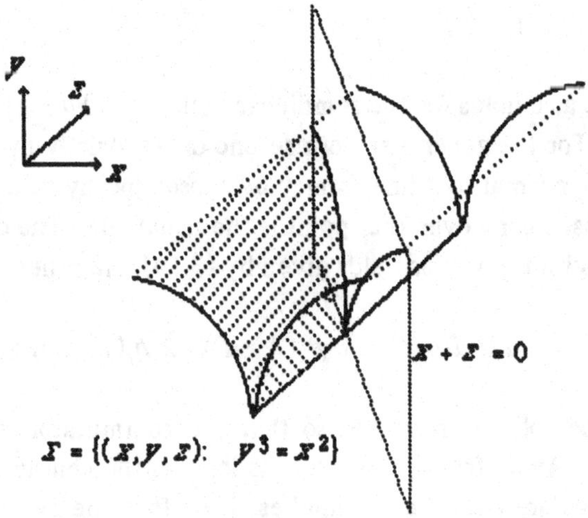

Figure 3.3

Away from the singular line $L = \{(0,0,z)\}$, Σ is a smooth characteristic hypersurface for $p_2(D)$. The vector fields in

$$\mathcal{M} = \{3x\partial_x + 2y\partial_y,\ 3y^2\partial_x + 2x\partial_y,\ \partial_z\}$$

are easily seen to be tangent to $\Sigma\backslash L$, and to generate a Lie algebra which away from L contains all smooth vector fields tangent to $\Sigma\backslash L$. (At L the picture is more complicated, since if D is any vector field and $a(x,y)$ is smooth and has support in $y \le 0$, then $a(x,y)D$ is tangent to $\Sigma\backslash L$.) We say that $u \in H^s_{loc}(\mathbf{R}^3)$ is conormal with respect to the cusp Σ if $M_1 \cdots M_j u \in$

$H^s_{loc}(\mathbf{R}^3)$ for all $M_1, \ldots, M_j \in \mathcal{M}$.

Theorem 3.19. *Suppose that* f *is smooth,* $u \in H^s_{loc}(\mathbf{R}^3)$, $s > 3/2$, *and* $p_2(D)u = (\partial_x\partial_z - \partial_y^2 + (9y/4)\partial_x^2)u = f(t,x,u)$. *If* u *is conormal with respect to the cusp* $\Sigma = \{(x,y,z)\colon\ y^3 = x^2\}$ *for* $x + z < 0$, *then* u *is conormal with respect to* Σ *on a neighborhood of the origin.*

Proof. Let $M_1 = 3x\partial_x + 2y\partial_y$, $M_2 = 3y^2\partial_x + 2x\partial_y$, and $M_3 = \partial_z$. The commutators of these vector fields with the operator satisfy

$$[p_2(D), M_1] = 4p_2(D) - \partial_x M_3,$$
$$[p_2(D), M_2] = 2\partial_y M_3 - \partial_x M_1 - 6\partial_x,$$
$$[p_2(D), M_3] = 0.$$

Therefore, the argument given in the proof of Theorem 3.6 again applies in this context. Q.E.D.

Solutions to nonlinear problems conormal in the past with respect to a smooth characteristic hypersurface which forms a swallowtail singularity in the future are in general expected to have singularities on the surface of the forward light cone emanating from the vertex of the swallowtail. The singularities are expected to be conormal with respect to the family consisting of the swallowtail and the surface of the forward light cone. The swallowtail singularity in the analytic setting is treated in Lebeau [41], [42], Delort [31].

In a sense, all of the conormal results described above are ones in which any interaction which takes place is really occuring in at most two dimensions. The conormal hypotheses essentially allow the problem to be reduced to one in a single space dimension. The only singularities present in the nonlinear case which are absent in the linear case are those described in Theorem 3.9, due to the presence of additional characteristic hypersurfaces in a problem of order greater than two. These nonlinear singularities have already been seen to occur in one space dimension (Theorem 2.1). Next to be considered is the case of a truly higher dimensional interaction, where new nonlinear conormal singularities exist even in a second order problem.

Chapter IV. Conormal Regularity after Nonlinear Interaction

When more than a pair of characteristic hypersurfaces carrying conormal singularities for a solution to a nonlinear problem intersect transversally in a lower dimensional manifold, new singularities can form, even for a second order equation. An example exhibiting this phenomenon was constructed in Rauch-Reed [60], with singularities as indicated in Figure 4.1. The solution u to a semilinear wave equation $\Box u = f(t,x,y,u)$ in two space dimensions is conormal in the past with respect to a triple of characteristic hyperplanes which intersect transversally at the origin. A new singularity is present at later times on the surface of the light cone over the origin. The nature of this singularity will be analyzed in detail below. Its presence is not surprising: any definition of conormal space for this geometry would include functions with wavefront sets over the origin having (τ,ξ,η) projections which include three linearly independent directions. An algebra of such functions would include u for which $WF(u) \supset \{(0,0,0,\tau,\xi,\eta)\colon\ (\tau,\xi,\eta) \neq 0\}$. Hörmander's Theorem would then allow for the propagation of such singularities onto the surface of the forward light cone. These are the only such singularities which arise, and they are also conormal, as shall also be established.

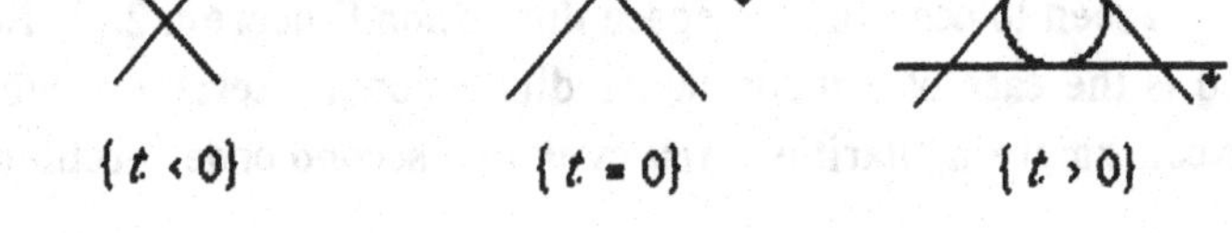

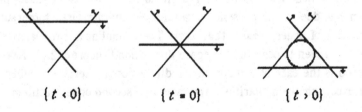

Figure 4.1

It is easiest to consider the example of a triple interaction in symmetric coordinates in $\mathbf{R}^3$, in which the characteristic hyperplanes are $\{x = 0\}$, $\{y = 0\}$, and $\{z = 0\}$, and the wave operator is

$$(4.1) \qquad \Box = \partial_x\partial_y + \partial_x\partial_z + \partial_y\partial_z .$$

This operator is strictly hyperbolic with respect to $t = x + y + z$, and the surface of the light cone is given by

$$\{\alpha = 2(xy + xz + yz) = 0\}.$$

Let the distance to the surface of the forward light cone $\{\alpha = 0\} \cap \{x > 0\}$ be denoted by $d(x,y,z)$. For $\beta \in C^\infty(\mathbf{R}^3)$, $\beta = 0$ for $t \leq -2$, $\beta = 1$ for $t \geq -1$, we will consider the solution to the following semilinear problem.

$$(4.2) \quad \Box u = \beta u^3 , \quad u = x_+^{\,k} + y_+^{\,k} + z_+^{\,k} \text{ for } x + y + z \leq -2.$$

Clearly, $x_+^{\,k} + y_+^{\,k} + z_+^{\,k}$ satisfies the linear homogeneous wave equation and is conormal with respect to $\{x = 0\}$,$\{y = 0\}$,$\{z = 0\}$, and $u \in H^s_{loc}(\mathbf{R}^3)$ for $s > 3/2$ as long as $k > 0$.

Theorem 4.1. *Let C denote the solid forward light cone over the origin for $\Box$. If u is the solution to* (4.2), *then $\partial_x^{k+1}\partial_y^{k+1}\partial_z^{k+1} u\big|_C$ has a singularity with principal term $c_0/d^{1/2}(x,y,z)$ on the complement of $\{x = 0\} \cup \{y = 0\} \cup \{z = 0\}$.*

As a consequence, it can be deduced that u is not piecewise smooth with respect to the family $\{x = 0\}$, $\{y = 0\}$, $\{z = 0\}$, $\{\alpha = 0\}$, even though it is piecewise smooth with respect to $\{x = 0\}$, $\{y = 0\}$, $\{z = 0\}$ in $\{x + y + z \leq -2\}$; see Beals [9]. (Piecewise smoothness of a function with respect to a collection of hypersurfaces means that the restriction of the function to each component of the complement of the collection extends to be C^∞ on the closure of the component.) For solutions to a semilinear strictly hyperbolic equation of any order, Rauch-Reed [61] establish that piecewise smoothness in the past with respect to a single characteristic hypersurface implies piecewise smoothness in the future. Moreover, for a second order semilinear equation or a "two-speed system", Rauch-Reed [62] prove that piecewise smoothness with respect to a pair of transversally intersecting characteristic hypersurfaces is preserved. On the other hand, Métivier-Rauch [51] demonstrate that piecewise smoothness is not in general preserved for semilinear solutions with singularities across a pair of characteristic hypersurfaces for higher order equations or systems, and give necessary and sufficient conditions on the transversal intersection for such smoothness to be preserved. These properties indicate why functions with conormal regularity may be considered a more appropriate category for solutions to nonlinear wave equations than functions which are piecewise smooth. Solutions with more restricted types of such regularity ("classical" conormal functions) have also been examined, for a single hypersurface in the semilinear case by Rauch-Reed [63] and in the quasilinear or fully nonlinear case by Piriou [55]. A pair of hypersurfaces for a semilinear classical conormal problem are treated in Nadir-Piriou [53].

Proof. Let $v = x_+^k + y_+^k + z_+^k$, so that

$$\begin{aligned} v^3 = {} & (x_+^{3k} + y_+^{3k} + z_+^{3k}) \\ & + 3(x_+^{2k}y_+^k + x_+^{2k}z_+^k + x_+^k y_+^{2k} + y_+^{2k}z_+^k + x_+^k z_+^{2k} + y_+^k z_+^{2k}) \\ & + 6(x_+^k y_+^k z_+^k) \\ = {} & s + p + 6(x_+^k y_+^k z_+^k). \end{aligned}$$

With E denoting the forward fundamental solution to $\Box$ starting at $t = -2$, as before we write

$$\begin{aligned} u &= v + E\beta u^3 = v + E\beta(v + E\beta u^3)^3 \\ &= v + E\beta(s + p + 6(x_+^k y_+^k z_+^k)) + E\beta(3v^2 E\beta u^3 + 3v(E\beta u^3) + (Ebu^3)^3). \end{aligned}$$

It follows easily from Theorem 3.3, Theorem 3.6, and superposition that $E\beta(s + p)$ is a sum of terms conormal with respect to the individual hypersurfaces $\{x = 0\}$, $\{y = 0\}$, $\{z = 0\}$ and pairs of these hypersurfaces. In particular, $v + E\beta(s + p)|_C$ is C^∞ away from $\{x = 0\} \cup \{y = 0\} \cup \{z = 0\}$. Since $\beta = 1$ on the support of $x_+^k y_+^k z_+^k$, it suffices to prove that $w = E(x_+^k y_+^k z_+^k)$ has a singularity of the specified type, and that

$$R = E\beta(3v^2 E\beta u^3 + 3v(E\beta u^3) + (E\beta u^3)^3)$$

is a strictly smoother remainder away from $\{x = 0\} \cup \{y = 0\} \cup \{z = 0\}$. It will be established below that

$$R \in H^{3k+7/2}{}_{loc}(C\backslash\{x=0\}\cup\{y=0\}\cup\{z=0\}).$$

Since $\square\,\partial_x^{k+1}\partial_y^{k+1}\partial_z^{k+1} w = \delta_0$, it follows that

$$\partial_x^{k+1}\partial_y^{k+1}\partial_z^{k+1} w = c_0/d^{1/2}(x,y,z)\,\chi_C(x,y,z).$$

(See, for example, Treves [71].) In particular, the principal nonlinear term in u is as claimed, and $w \notin H^{3k+3}{}_{loc}(C\backslash\{x=0\}\cup\{y=0\}\cup\{z=0\})$. Q.E.D.

In order to measure the regularity of the remainder term in the above argument, we define Sobolev spaces appropriate to the geometry of this problem. Since the solutions $x_+{}^k$, $y_+{}^k$, and $z_+{}^k$ to the linear wave equation have different regularity with respect to the three coordinate directions, it is natural to take such smoothness into account. Moreover, since for all $\varepsilon > 0$, $x_+{}^k \in H^{(k+1/2+\varepsilon)}{}_{loc}(\mathbf{R}^3)$, it is simplest to work in spaces of functions which just fail to be in $H^s{}_{loc}(\mathbf{R}^3)$.

Definition 4.2. For $k_1, k_2, k_3 \geq 0$, $u \in H^{s-;k_1,k_2,k_3}$ means that

$$\langle\xi\rangle^{k_1}\langle\eta\rangle^{k_2}\langle\zeta\rangle^{k_3}\langle(\xi,\eta,\zeta)\rangle^{s-\varepsilon}(\varphi u)^\wedge(\xi,\eta,\zeta) \in L^2(\mathbf{R}^3),$$

for any $\varepsilon > 0$ and $\varphi \in C^\infty{}_{com}(\mathbf{R}^3)$.

The proof of Lemma 1.3 may easily be adapted to establish the algebra properties of these spaces.

Lemma 4.3. *If* $s \geq 0$ *and* $\min\{s+k_j > 1/2\}$, *then* $H^{s-;k_1,k_2,k_3}$ *is an algebra. Moreover, if* $a_j \geq 0$ *and* $a_1 + a_2 + a_3 = 1$, *then* $H^{s+1-;k_1,k_2,k_3} \subset H^{s-;k_1+a_1,k_2+a_2,k_3+a_3}$.

In addition, since $\square$ commutes with each of the vector fields $\partial_x, \partial_y, \partial_z$, a simple commutator argument and the linear energy inequality imply that

(4.3) $\quad E$ maps $H^{s-;k_1,k_2,k_3}$ into $H^{s+1-;k_1,k_2,k_3}$.

If u satisifies (4.2), then for any $\varepsilon > 0$, $u \in H^{k+1/2-\varepsilon}{}_{loc}(\{t < -2\})$, so the linear energy inequality implies that

$$u \in H^{k+1/2-\epsilon}_{loc}(\mathbf{R}^3) \subset H^{0-;(k+1/2)/3,(k+1/2)/3,(k+1/2)/3}.$$

Since $v = x_+^k + y_+^k + z_+^k$ satisfies

$$v \in H^{0-;k+1/2,\infty,\infty} + H^{0-;\infty,k+1/2,\infty} + H^{0-;\infty,\infty,k+1/2},$$

the expression $u = v + E\beta u^3$, Lemma 4.2, (4.3), and induction easily imply that $u \in H^{0-;k+1/2,k+1/2,k+1/2}$ and $E\beta u^3 \in H^{1-;k+1/2,k+1/2,k+1/2}$. Since

$$v^2 \in H^{0-;k+1/2,k+1/2,\infty} + H^{0-;k+1/2,\infty,k+1/2} + H^{0-;\infty,k+1/2,k+1/2},$$

it follows from Lemma 4.2 that

$$\begin{aligned} 3v^2 E\beta u^3 \in\ & H^{0-;k+1/2,k+1/2,k+3/2} + H^{0-;k+1/2,k+3/2,k+1/2} \\ & + H^{0-;k+3/2,k+1/2,k+1/2}, \end{aligned}$$

and then from (4.3) that

$$\begin{aligned} R &= E\beta(3v^2 E\beta u^3 + 3v(E\beta u^3) + (E\beta u^3)^3) \\ &\in H^{1-;k+1/2,k+1/2,k+3/2} + H^{1-;k+1/2,k+3/2,k+1/2} \\ &\quad + H^{1-;k+3/2,k+1/2,k+1/2}. \end{aligned}$$

It will be established in Theorem 4.8 that u is conormal with respect to the family $\{x=0\}, \{y=0\}, \{z=0\}, \{\alpha=0\}$, so in particular, the wavefront set of $u|_{C\setminus\{x=0\}\cup\{y=0\}\cup\{z=0\}}$ is near $N^*(\{\alpha=0\}\setminus\{x=0\}\cup\{y=0\}\cup\{z=0\})$. On the (ξ,η,ζ) projection of this set, $\langle\xi\rangle, \langle\eta\rangle, \langle\zeta\rangle$, and $\langle\langle\xi,\eta,\zeta\rangle\rangle$ are all comparable. Consequently, $R \in H^{3k+7/2}_{loc}(C\setminus\{x=0\}\cup\{y=0\}\cup\{z=0\})$, completing the proof of Theorem 4.1.

In general, after each triple intersection of characteristic hypersurfaces carrying conormal singularities for a solution to a nonlinear wave equation in more than one space dimension, an additional nonlinear singularity will be present at later times on the surface of the light cone over the triple intersection point; see Holt [34]. Such singularities can quickly accumulate. The analogue in the case of one space dimension for higher order equations is considered in Rauch-Reed [59]. For a fourth order operator with principal part $(\partial_t - \partial_x)(\partial_t - 2\partial_x)(\partial_t + \partial_x)(\partial_t + 2\partial_x)$ and for appropriate nonlinearity, a solution exists with singularities as in Figure 4.2. Although only finitely many singularities are present initially, after finite time the solution has singular support with nonempty interior. Of course, Theorem 1.11 implies that later generation singularities are weaker than the ones producing them, so that given any finite measure of regularity (for instance $H^r_{loc}(\mathbb{R}^2)$, $r < \infty$) the set of singularities of at most that strength will be a locally finite union of characteristic curves.

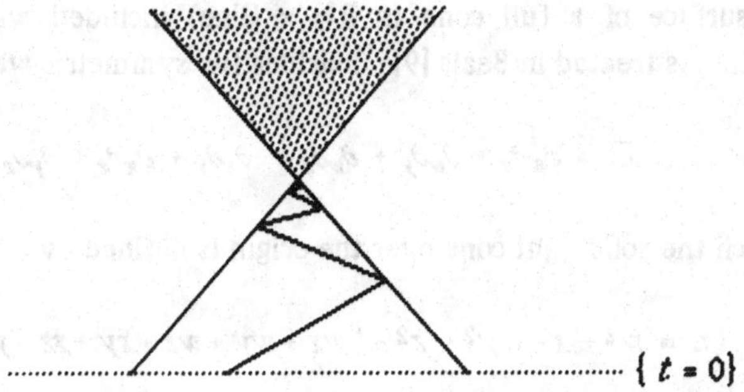

Figure 4.2

For two space dimensions and solutions to $\Box u = f(t,x,u)$, Melrose-Ritter [47] consider solutions initially conormal with respect to a finite collection of points and establish instances for which the expected locus of singularities has nonempty interior at later times. The mimimal such configuration, having four initial singularities, is considered in Sà Baretto-Melrose [66]. Triple interactions include the intersection of a tangent pair (a characteristic hypersurface and the surface of the light cone arising from an earlier interaction) with an additional transversal hypersurface, as in Figure 4.3.

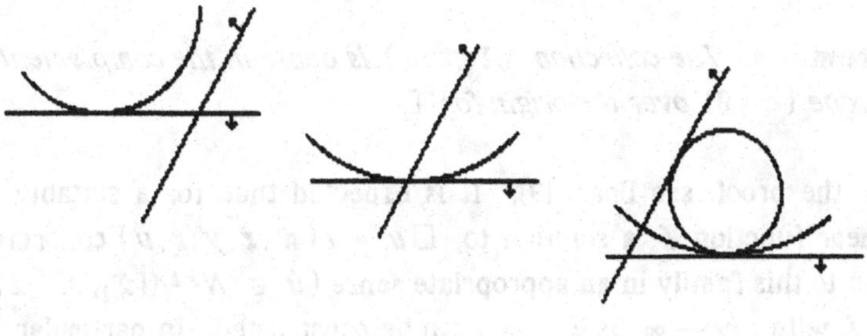

Figure 4.3

In three space dimensions, the corresponding geometry for the interaction of conormal singularities for a semilinear wave equation is more complicated. For example, consider the multiple interaction due to the simultaneous transversal intersection of four characteristic hypersurfaces. In addition to the surface of the light cone over the point of quadruple interaction, from every triple interaction, now occuring along lines in $\mathbf{R}^4$, one expects nonlinear singularities on hypersurfaces of the form (surface of a half-cone in $\mathbf{R}^3$) × (line). An example treating singularities on hypersurfaces of the form (surface of a full cone in $\mathbf{R}^3$) × (line) included with every triple interaction is treated in Beals [9]. Consider the symmetric wave operator

$$\square = \partial_w\partial_x + \partial_w\partial_y + \partial_w\partial_z + \partial_x\partial_y + \partial_x\partial_z + \partial_y\partial_z ,$$

for which the solid light cone over the origin is defined by

$$\{\alpha = w^2 + x^2 + y^2 + z^2 - (wx + wy + wz + xy + xz + yz) \leq 0\},$$

and start with the characteristic hyperplanes $\{w = 0\}, \{x = 0\}, \{y = 0\}, \{z = 0\}$. From each triple intersection along a line is generated the corresponding characteristic hypersurface formed by the (t, x) projection of the union of the null bicharacteristics for $\square$ over the line. (For example, the set generated over $\{x = y = z = 0\}$ is $\{2(xy + xz + yz) - (x^2 + y^2 + z^2) = 0\}$.) Let $\{\Sigma_1, \ldots, \Sigma_{m_1}\}$ be the collection consisting of the original hypersurfaces and the four hypersurfaces generated at this first stage. Given $\{\Sigma_1, \ldots, \Sigma_{m_j}\}$, let $\{\Sigma_1, \ldots, \Sigma_{m_{j+1}}\}$ be the union of $\{\Sigma_1, \ldots, \Sigma_{m_j}\}$ and all of the characteristic hypersurfaces generated over lines of triple intersection of hypersurfaces in $\{\Sigma_1, \ldots, \Sigma_{m_j}\}$. Each of these sets is a conic subset the complement of the light cone $\{\alpha < 0\}$. Let their union be denoted by $\{\Sigma_1, \ldots\}$.

Theorem 4.4. *The collection* $\{\Sigma_1, \ldots\}$ *is dense in the complement of the light cone* $\{\alpha < 0\}$ *over the origin for* $\square$.

For the proof, see Beals [9]. It is expected that, for a suitably chosen nonlinear function f, a solution to $\square u = f(w,x,y,z,u)$ conormal with respect to this family in an appropriate sense ($u \in N^{s,k_j}(\{\Sigma_1, \ldots, \Sigma_{m_j}\})$ for each j, with $k_j \to \infty$ as $j \to \infty$) can be constructed. In particular, such a function would in general have *sing supp*$(u) \supset \{\alpha \leq 0\}$. On the other hand, since all of the characteristic hypersurfaces considered above are conic subsets of $\mathbf{R}^4$, any notion of conormal regularity will require that u not lose regularity when differentiated with respect to the radial vector field $w\partial_w + x\partial_x + y\partial_y + z\partial_z$. It follows from Theorem 3.17 that such a solution will necessarily be smooth on $\{\alpha > 0\}$, the interior of the solid light cone over the origin.

The interaction of singularities conormal in the past with respect to a triple of hyperplanes characteristic for $\Box$, which intersect transversally at the origin, is the simplest case of a conormal regularity problem in which straightforward commutator arguments will not work. The example given in Theorem 4.1 of a solution semilinear wave equation $\Box u = f(t,x,y,u)$ in two space dimensions has a singularity on the surface of the forward light cone over the origin. The family of vector fields tangent to the three original hyperplanes and the surface of the cone (or the family of classical first order pseudodifferential operators with principal symbols vanishing on the corresponding conormal bundles) does not satisfy the appropriate commutation property as in the proof of Theorem 3.6. Nevertheless, conormal regularity is preserved: the solution remains conormal with respect to this family of surfaces in the future. This result was first established by Melrose-Ritter [46] and by Bony [18]. Melrose and Ritter followed the procedure of blowing up the singularity at the origin (that is, using polar coordinates in $\mathbf{R}^3$), analyzing the singularity in the lifted coordinates, where microlocally the pertinent manifolds are separated, and proving a new weighted energy inequality for the linear inhomogeneous wave equation. Bony used the technique of second microlocalization, which had been developed in the analytic case by Laurent [39], in order to localize simultaneously in regions where $x/|x|$ and $\xi/|\xi|$ do not vary too quickly. This argument is refined in Bony [19] into a new symbolic calculus tailored to the second microlocal analysis of the problem. The higher order microlocal calculus is treated in Bony-Lerner [20].

A different proof is given in Beals [10]. First, a commutator argument is used to obtain regularity with respect to most of the necessary vector fields. Then, regularity with respect to the remaining vector fields is obtained by use of the equation itself and the earlier vector fields. Only the usual linear energy estimate is used, and no pseudodifferential operators are necessary. We consider first the flat case, in symmetric coordinates, with the characteristic hyperplanes given by $\{x = 0\}$, $\{y = 0\}$, and $\{z = 0\}$, the wave operator given by (4.1), and the surface of the light cone given by

$$\{\alpha \equiv 2(xy + xz + yz) = 0\}.$$

It will be necessary to localize in cones around the origin, using functions $\chi \in C^\infty(\mathbf{R}^3\backslash 0)$ which are homogeneous of degree zero. Then $\chi \in H^{3/2-\epsilon}{}_{loc}(\mathbf{R}^3)$ for every $\epsilon > 0$ for such functions, so it is natural to work in the following space.

Definition 4.5. $H^{n/2-} = \bigcap_{\epsilon>0} H^{n/2-\epsilon}{}_{loc}(\mathbf{R}^n)$.

These spaces are polynomial algebras, as follows from (1.5). Moreover, for conormal arguments we will in general apply nonlinear functions to elements u of $H^{n/2+\epsilon}{}_{loc}(\mathbf{R}^n)$ and multiply by functions in $H^{n/2-}$. In this case, any smooth function of u may be used, rather than just a polynomial.

Lemma 4.6. *Let $\mathcal{M}$ be a Lie algebra of vector fields on $\mathbf{R}^n$ with coefficients in $C^\infty(\mathbf{R}^n\backslash 0)$ which are homogeneous of degree one. Suppose that $u \in H^{n/2+\epsilon}{}_{loc}(\mathbf{R}^n)$ and $M_1 \cdots M_j u \in H^{n/2-}$ for $M_1, \ldots, M_j \in \mathcal{M}$, $j \leq k$. If $f(x,v) \in C^\infty$, then $M_1 \cdots M_j f(x,u) \in H^{n/2-}$ for $M_1, \ldots, M_j \in \mathcal{M}$, $j \leq k$.*

Proof. If $\chi \in C^\infty(\mathbf{R}^n\backslash 0)$ is homogeneous of degree zero and $M \in \mathcal{M}$, then $M\chi \in C^\infty(\mathbf{R}^n\backslash 0)$ is homogeneous of degree zero. It follows immediately from Schauder's lemma, the chain rule, and (1.5) that $M_1 \cdots M_j f(x,u) \in H^{n/2-}$, because $M_1 \cdots M_j f(x,u)$ may be written as a polynomial in elements of $H^{n/2-}$. Q.E.D.

If K is a cone with vertex at the origin, vector fields with coefficients in $C^\infty(K\backslash 0)$ which are homogeneous of degree one will be referred to as admissible on K. We use such vector fields to define conormal regularity with respect to the family of hypersurfaces under consideration.

Definition 4.7. Let $\mathcal{M}$ be the Lie algebra of vector fields with coefficients in $C^\infty(\mathbf{R}^3\backslash 0)$ which are homogeneous of degree one and simultaneously tangent to all of the surfaces $\{x = 0\}, \{y = 0\}, \{z = 0\}$, and $\{\alpha = 0\}$. Then $u \in H^{3/2-}$ is said to be conormal with respect to the family $\{x = 0\}, \{y = 0\}, \{z = 0\}, \{\alpha = 0\}$ if $M_1 \cdots M_j u \in H^{3/2-}$ for all $M_1, \cdots, M_j \in \mathcal{M}$. If this property holds for $j \leq k$, u is said to be conormal of degree k, written

$$u \in N^{3/2-,k}(\{x = 0\}, \{y = 0\}, \{z = 0\}, \{\alpha = 0\}).$$

It easily follows by use of a smooth partition of unity on S^2 that u is conormal if and only if the following conditions hold. For any $\chi \in C^\infty(\mathbf{R}^n \backslash 0)$ which is homogeneous of degree zero and for which $supp(\chi)\backslash 0$ meets at most a pair of the hypersurfaces, $M_1 \cdots M_j \chi u \in H^{3/2-}$ for all admissible vector fields $M_1, \ldots, M_j$ on $supp(\chi)$ which are simultaneously tangent to all of the surfaces meeting $supp(\chi)\backslash 0$. By Lemma 4.6,

(4.4) $N^{3/2-,k}(\{x=0\},\{y=0\},\{z=0\},\{\alpha=0\})$ is a polynomial algebra, and $H^{3/2+\epsilon}_{loc}(\mathbf{R}^3) \cap N^{3/2-,k}(\{x=0\},\{y=0\},\{z=0\},\{\alpha=0\})$ is preserved under the action of smooth functions.

The regularity result in the flat case of the conormal triple interaction may now be stated. In the case $k = \infty$, the singularities are in general supported on the sets indicated in Figure 4.3.

Theorem 4.8. *Let $u \in H^{3/2+\epsilon}_{loc}(\mathbf{R}^3)$, let f be smooth, let $\Box$ be as in (4.1), and suppose that $\Box u = f(x,y,z,u)$. If $u \in N^{3/2-,k}(\{x=0\},\{y=0\},\{z=0\},\{\alpha=0\})$ for $x+y+z < t_0$, then $u \in N^{3/2-,k}(\{x=0\},\{y=0\},\{z=0\},\{\alpha=0\})$.*

Proof. By the pairwise interaction result (Theorem 3.6) and superposition, it can be assumed that $f = 0$ for $x+y+z \leq t_0$. Therefore, if E is the forward fundamental solution for $\Box$ starting at t_0, we may again write $u = v + Ef(x,y,z,u)$, with v a solution of $\Box v = 0$. Since, by linearity, v is a sum of terms conormal with respect to the pairs

$$(\{x=0\},\{y=0\}),\ (\{x=0\},\{z=0\}),(\{y=0\},\{z=0\}),$$

it easily follows that $v \in N^{3/2-,k}(\{x=0\},\{y=0\},\{z=0\},\{\alpha=0\})$. Consequently, it will be assumed without loss of generality that $u = Ef(x,y,z,u)$. Notice that, if M is a smooth vector field satisfying the commutation relation $[\Box, M] = c\Box$ for some constant c, then

(4.5) $\quad MEg = E(M+c)g$ for g supported in $x+y+z \geq t_0$.

A collection of smooth vector fields with this property will be considered; each will be tangent to at least three of the four hypersurfaces in question. A conic partition of unity will reduce the argument to the use of these operators, and to an additional collection of vector fields which may be obtained by use of the equation. The proof is completed in the series of steps which follow.

Definition 4.9. Set $M_0 = x\partial_x + y\partial_y + z\partial_z$, $M_1 = (y - z)\partial_x + y\partial_y - z\partial_z$, $M_2 = -x\partial_x + (z - x)\partial_y + z\partial_z$, and $M_3 = x\partial_x - y\partial_y + (x - y)\partial_z$.

The radial vector field M_0 is tangent to all hyperplanes through the origin, as well as the surface of the light cone $\{\alpha = 0\}$. Two of the vector fields M_1, M_2, and M_3 are tangent to each of the hyperplanes $\{x = 0\}$, $\{y = 0\}$, $\{z = 0\}$. Moreover, the third vector field may be expressed in terms of the other two (where the coefficients do not vanish), since

$$xM_1 + yM_2 + zM_3 = 0 .$$

Furthermore, all of these vector fields are tangent to the hypersurface $\{\alpha = 0\}$, and, in addition, are well behaved with respect to the d'Alembertian $\Box$. The following properties are easily verified.

Lemma 4.10. *a) For each* M_i, $i = 1, 2, 3$, *the commutator with* $\Box$ *satisfies* $[\Box, M_i] = 0$. *For the radial vector field*, $[\Box, M_0] = 2\Box$.
b) $M_i\alpha = 0$ *for* $i = 1, 2, 3$; $M_0\alpha = 2\alpha$.

On cones near the intersection of the surface of the light cone with any one of the coordinate hyperplanes, these four vector fields are sufficient to generate all admissible vector fields simultaneously tangent to the hypersurfaces in question. For the straightforward verification of the following properties, see Beals [10].

Lemma 4.11. *Let* K *be an open conic neighborhood of* $\{\alpha = 0\} \cap \{x = 0\}$. *The collection* $\{M_0, M_2, M_3\}$ *generates all admissible vector fields on* K *which are simultaneously tangent to* $\{\alpha = 0\}$ *and* $\{x = 0\}$. *Analogous statements hold for* $\{M_0, M_1, M_3\}$ *with respect to* $\{\alpha = 0\}$ *and* $\{y = 0\}$, *and for* $\{M_0, M_1, M_2\}$ *with respect to* $\{\alpha = 0\}$ *and* $\{z = 0\}$.

A commutator argument may now be used on cones over the origin. Let $\chi'_1, \chi'_{1,2} \in C^\infty(\mathbf{R}^n \setminus 0)$ be homogeneous of degree zero and have support on small conic neighborhoods of $\{\alpha = 0\} \cap \{x = 0\}$ and $\{x = 0\} \cap \{y = 0\}$

respectively. See Figure 4.4 for a time slice.

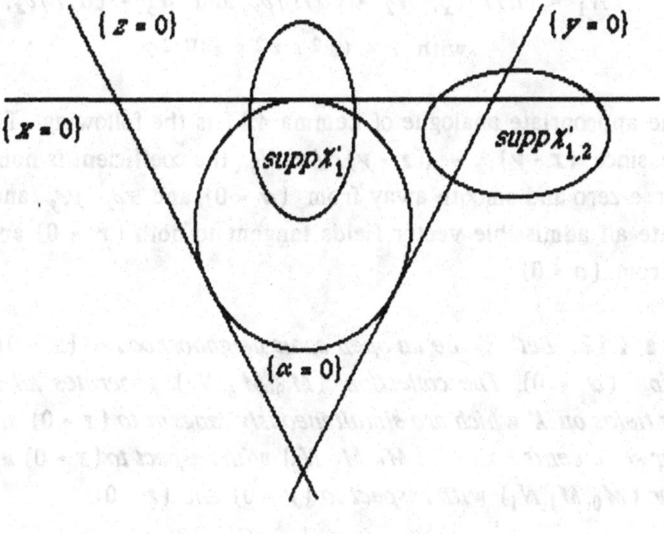

Figure 4.4

By symmetry, it is enough to establish conormal regularity for

$$u_1 = E\chi'_1 f(x,y,z,u),$$
$$u_{1,2} = E\chi'_{1,2} f(x,y,z,u).$$

(The terms involving $E\chi f(x,y,z,u)$ for χ with support intersecting fewer of the hypersurfaces may be handled as in the argument for u_1, using all four of the vector fields in Definition 4.9. In particular, the terms involving those χ having nontrivial intersection with $\{t = 0\}$ fall into this category, and thus we treat the configuration shown in the time slice in Figure 4.4 as generic.) Suppose inductively that $u \in N^{3/2-,j}(\{x=0\},\{y=0\},\{z=0\},\{\alpha=0\})$ for $0 < j < k$. Then from (4.4), (4.5), Lemma 4.10, and the linear energy estimate (which implies that E maps functions in $H^{3/2-}$ which vanish in the past into $H^{3/2-}$),

$$M_0^{\beta_0} M_2^{\beta_2} M_3^{\beta_3} u_1 \in H^{3/2-} \text{ for } |\beta| \le j+1. \tag{4.6}$$

On the other hand, only the two vector fields M_0 and M_3 are tangent to the hypersurfaces in question on $supp(\chi'_{1,2})$. Therefore, we introduce a further collection of operators which do not commute with $\Box$ in order to treat the regularity of $u_{1,2}$.

$$N_1 = (\alpha/r)\partial_x,\ N_2 = (\alpha/r)\partial_y, \text{ and } N_3 = (\alpha/r)\partial_z, \quad \text{with } r = (x^2+y^2+z^2)^{1/2}. \tag{4.7}$$

The appropriate analogue of Lemma 4.11 is the following. The proof is simple, since $(x-y)\partial_z = \{(x-y)r/\alpha\}N_3$, the coefficient is homogeneous of degree zero and smooth away from $\{\alpha = 0\}$, and $x\partial_x$, $y\partial_y$, and $(\alpha/r)\partial_z$ generate all admissible vector fields tangent to both $\{x = 0\}$ and $\{y = 0\}$ away from $\{\alpha = 0\}$.

Lemma 4.12. *Let K be an open conic neighborhood of $\{x = 0\} \cap \{y = 0\}$ avoiding $\{\alpha = 0\}$. The collection $\{M_0, M_3, N_3\}$ generates all admissible vector fields on K which are simultaneously tangent to $\{x = 0\}$ and $\{y = 0\}$. Similar statements hold for $\{M_0, M_2, N_2\}$ with respect to $\{x = 0\}$ and $\{z = 0\}$, and for $\{M_0, M_1, N_1\}$ with respect to $\{y = 0\}$ and $\{z = 0\}$.*

The substitute for a commutator argument is the following property, which may be established by expanding the expression for $xy\Box$ in terms of M_0, M_3, and N_3.

Lemma 4.13. *Modulo terms of first order in M_0, M_3, and N_3,*

$$(N_3)^2 = -4xy\Box + (M_0)^2 - (M_3)^2 + 2(x+y-z)N_3M_0.$$

Again, suppose that $u \in N^{3/2-,j}(\{x = 0\},\{y = 0\},\{z = 0\},\{\alpha = 0\})$ for $0 < j < k$. If $\beta_1 = 0$ or 1, the following inclusion is a result of (4.6) or the property that E maps $H^{3/2-}$ into $H^{5/2-}$ when acting on functions which are identically zero in the far past. If $\beta_1 \geq 2$, Lemma 4.13 and the equation $\Box u_{1,2} = \chi'_{1,2}f(u)$ may be used repeatedly to reduce to the case of $\beta_1 = 1$.

$$M_0^{\beta_0} N_3^{\beta_1} M_3^{\beta_3}\, u_{1,2} \in H^{3/2-} \text{ for } |\beta| \leq j+1. \tag{4.8}$$

The proof of Theorem 4.8 is completed by decomposing u_1 and $u_{1,2}$ into pieces supported on small cones around the origin and expressing the appropriate vector fields on each of the cones in terms of those given in (4.6) or (4.8). Let $\chi_1, \chi_3, \chi_{1,2}, \chi_{2,3} \in C^\infty(\mathbf{R}^3\backslash 0)$ be homogeneous of degree zero and have support on small conic neighborhoods as indicated in Figure 4.5

(where a time slice is shown). By symmetry, it is enough to establish regularity for $\chi_1 u_1, \chi_3 u_1, \chi_{1,2} u_1, \chi_{2,3} u_1, \chi_1 u_{1,2}, \chi_3 u_{1,2}, \chi_{1,2} u_{1,2}$, and $\chi_{2,3} u_{1,2}$, since the remaining terms (with fewer hypersurfaces) are even easier to handle. It follows immediately from (4.6), Lemma 4.11, (4.8), and Lemma 4.12 that $\chi_1 u_1, \chi_{1,2} u_{1,2} \in N^{3/2-,j+1}(\{x=0\},\{y=0\},\{z=0\},\{\alpha=0\})$.

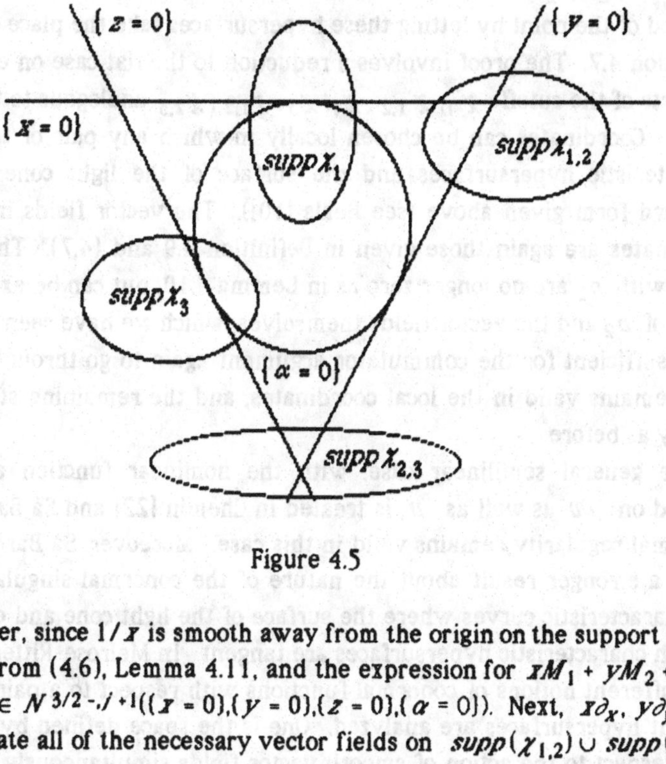

Figure 4.5

Moreover, since $1/x$ is smooth away from the origin on the support of χ_3, it follows from (4.6), Lemma 4.11, and the expression for $xM_1 + yM_2 + zM_3$ that $\chi_3 u_1 \in N^{3/2-,j+1}(\{x=0\},\{y=0\},\{z=0\},\{\alpha=0\})$. Next, $x\partial_x$, $y\partial_y$, and $z\partial_z$ generate all of the necessary vector fields on $supp(\chi_{1,2}) \cup supp(\chi_{2,3})$, and, for instance, we may write $x\partial_x = (r^2/\alpha)((\alpha x/r^2)\partial_x)$, with coefficient which is smooth away from the origin on the supports of $\chi_{1,2}$ and $\chi_{2,3}$. Similar expressions hold for $y\partial_y$ and $z\partial_z$. But $(\alpha x/r^2)\partial_x, (\alpha y/r^2)\partial_y$, and $(\alpha z/r^2)\partial_z$ are tangent to all four of the hypersurfaces under consideration, so Lemma 4.11 implies that they may be expressed in terms of M_0, M_2, and M_3. Hence from (4.6), $\chi_{1,2} u_1, \chi_{2,3} u_1 \in N^{3/2-,j+1}(\{x=0\},\{y=0\},\{z=0\},\{\alpha=0\})$. Similarly, $\chi_{2,3} u_{1,2} \in N^{3/2-,j+1}(\{x=0\},\{y=0\},\{z=0\},\{\alpha=0\})$ from Lemma 4.13. Finally, on $supp(\chi_3)$ the vector fields M_1 and M_2 may be expressed in terms of M_0, $(z/r)M_3$, and $(z/r)N_3$, with coefficients which are smooth away from the origin, while M_2 may be expressed in terms of M_0, M_3, and $(x/r)N_3$ on $supp(\chi_1)$. Therefore, (4.8) and Lemma 4.11 yield that $\chi_3 u_{1,2}$, $\chi_1 u_{1,2} \in N^{3/2-,j+1}(\{x=0\},\{y=0\},\{z=0\},\{\alpha=0\})$, thus completing the inductive step.

Theorem 4.8 remains valid with any second order strictly hyperbolic operator p_2 in place of $\Box$, and any triple of smooth transversally intersecting characteristic hypersurfaces. In this case the surface of the light cone over the triple intersection is the projection of the union of all of the null bicharacteristics for p_2 over the point. The conormal space is defined on a neighborhood of the point by letting these hypersurfaces take the place of those in Definition 4.7. The proof involves a reduction to the flat case on each of the supports of the cutoffs $\chi'_1, \chi'_{1,2}, \chi_1, \chi_3, \chi_{1,2}, \chi_{2,3}$ analogous to those used above. Coordinates can be chosen locally in which any pair of the smooth characteristic hypersurfaces and the surface of the light cone have the standard form given above (see Beals [10]). The vector fields in the local coordinates are again those given in Definition 4.9 and (4.7). The commutators with p_2 are no longer zero as in Lemma 4.10, but can be expressed in terms of p_2 and the vector fields themselves, which we have seen in Chapter III is sufficient for the commutator argument again to go through. Lemma 4.13 remains valid in the local coordinates, and the remaining steps follow exactly as before.

The general semilinear case, with the nonlinear function allowed to depend on Du as well as u, is treated in Chemin [22] and Sà Barreto [65]. Conormal regularity remains valid in this case. Moreover, Sà Barreto establishes a stronger result about the nature of the conormal singularity along the characteristic curves where the surface of the light cone and each of the smooth characteristic hypersurfaces are tangent. In Melrose-Ritter [47], several different notions of conormal functions with respect to a pair of simply tangent hypersurfaces are analyzed. One is the space defined by regularity with respect to the action of smooth vector fields simultaneously tangent to both hypersurfaces, which is the type of regularity present after the triple interaction proved in Melrose-Ritter [46] and Bony [18] (and above). However, it is conjectured that the subsequent interaction of a tangent pair of conormal singularities with an additional characteristic surface (as in Figure 4.3) would allow for singularities on a set with nonempty interior, if the tangent pair had only this conormal regularity. Melrose and Ritter consider the following stronger definition of conormal for a tangent pair. For hypersurfaces Σ_1 and Σ_2, with tangential intersection Δ, f is said to be conormal if $f = f_1 + f_2$, with each f_j conormal with respect to the pair (Σ_j, Δ). (The latter statement refers to regularity with respect to the action of all smooth vector simultaneously tangent to Σ_j and Δ.) This space is strictly smaller than that given by the earlier definition. In Melrose-Ritter [47] it is proved that this stronger regularity is preserved for solutions of second order semi-linear strictly hyperbolic problems if no other interactions occur. In Sà Barreto [65] it is shown that this stronger regularity is actually present in solutions after a transversal triple interaction, and moreover, strong conormal regularity is preserved after an interaction involving conormal singularities on a tangent pair and an additional characteristic hypersurface.

We consider the notion of strong conormal regularity for a pair of simply tangent hypersurfaces, away from the singular point corresponding to the point of triple interaction in Theorem 4.8, and treat a model case. For notational simplicity we treat the three dimensional problem in which the hypersurfaces are given by $\{y = 0\}$ and $\{y = x^2/4\}$. These hypersurfaces are characteristic for the strictly hyperbolic operator

$$p_2(D) = \partial_y\partial_z - \partial_x^2 + y\partial_y^2. \tag{4.9}$$

It is convenient to measure conormal regularity using the spaces of bounded functions which remain in $L^2{}_{loc}(\mathbf{R}^3)$ when acted upon by appropriate vector fields, as in Melrose-Ritter [47]. The following definition is equivalent to that given in [47].

Definition 4.14. Let $\mathcal{M}_1$ be the Lie algebra of smooth vector fields which are simultaneously tangent to $\{y = 0\}$ and $\{x = 0, y = 0\}$, and let $\mathcal{M}_2$ be the Lie algebra of smooth vector fields which are simultaneously tangent to $\{y = x^2/4\}$ and $\{x = 0, y = 0\}$. Then $u \in L^\infty{}_{loc}(\mathbf{R}^3)$ is said to be strongly conormal with respect to the tangent pair $\{x = 0\},\{x = z^2/4\}$ if for $\chi \in C^\infty(\mathbf{R})$, $\chi(r) = 1$, $r \leq 1/4$, $\chi(r) = 0$, $r \geq 1/2$, we have

$$M_1\cdots M_j\chi(4y/x^2)u \in L^2{}_{loc}(\mathbf{R}^3) \text{ for all } M_1,\ldots, M_j \in \mathcal{M}_1,$$
$$M_1\cdots M_j(1-\chi(4y/x^2))u \in L^2{}_{loc}(\mathbf{R}^3) \text{ for all } M_1,\ldots, M_j \in \mathcal{M}_2.$$

A set of generators of $\mathcal{M}_1$ is given by $\{x\partial_x, y\partial_y, y\partial_x, \partial_z\}$, while a change of variables easily yields that $\mathcal{M}_2$ is spanned by $\{x\partial_x + 2y\partial_y, (y-x^2/4)\partial_x, (y-x^2/4)\partial_y, \partial_z\}$. Since $\{x\partial_x + 2y\partial_y, (y-x^2/4)\partial_x, y\partial_y, \partial_z\}$ is also a set of generators of $\mathcal{M}_1$, we set

$$M_0 = x\partial_x + 2y\partial_y,\ M_1 = (y - x^2/4)\partial_x,\ M_2 = \partial_z, \tag{4.10}$$
$$N_1 = y\partial_y,\ N_2 = (y - x^2/4)\partial_y.$$

Lemma 4.15. *If u is strongly conormal with respect to the tangent pair $\{y = 0\}$, $\{y = x^2/4\}$, and f is smooth, then $f(u)$ is strongly conormal with respect to $\{y = 0\}$, $\{y = x^2/4\}$.*

Proof. Let $\chi \in C^\infty(\mathbf{R})$, be constant near $\pm\infty$. It is easily checked that, for any choice M of the vector fields given in (4.10), there are smooth functions $\alpha(x,y,z)$ and $\chi_1 \in C^\infty(\mathbf{R})$, with χ_1 constant near $\pm\infty$, such that

$$M(\chi(4y/x^2)) = \alpha(x,y,z)\chi_1(4y/x^2).$$

(For example, $N_1(\chi(4y/x^2)) = -(x/2)\chi_1(4y/x^2)$, with $\chi_1(r) = r^2\chi'(r)$.) If χ is chosen as in Definition 4.14, the chain rule therefore implies that

$$M_1 \cdots M_j\chi(4y/x^2)f(u) \in L^2{}_{loc}(\mathbf{R}^3) \text{ for } M_1, \ldots, M_j \in \mathcal{M}_1$$

if $\chi_1(4y/x^2)M_1 \cdots M_j f(u) \in L^2{}_{loc}(\mathbf{R}^3)$ for all $\chi_1 \in C^\infty(\mathbf{R})$ with support where $r \le 1/2$. Similarly,

$$M_1 \cdots M_j(1-\chi(4y/x^2))f(u) \in L^2{}_{loc}(\mathbf{R}^3) \text{ for } M_1, \ldots, M_j \in \mathcal{M}_2$$

if $\chi_2(4y/x^2)M_1 \cdots M_j f(u) \in L^2{}_{loc}(\mathbf{R}^3)$ for all $\chi_2 \in C^\infty(\mathbf{R})$ with support where $r \ge 1/4$.

By assumption, and from the preceding calculations, $u \in L^\infty{}_{loc}(\mathbf{R}^3)$ and

$$\chi(4y/x^2)M_1 \cdots M_j u \in L^2{}_{loc}(\mathbf{R}^3) \text{ for } M_1, \ldots, M_j \in \mathcal{M}_1.$$

The proof of Lemma 3.10, with $\mathcal{M}_1$ in place of the collection of all first order derivatives, yields that $\chi_1(4y/x^2)M_1 \cdots M_i u \in L^{2j/i}(\mathbf{R}^n)$ for $M_1, \ldots, M_i \in \mathcal{M}_1$, $1 \le i \le j$. As in the proof of Corollary 3.11, it follows from Hölder's inequality that $\chi_1(4y/x^2)M_1 \cdots M_j f(u) \in L^2{}_{loc}(\mathbf{R}^3)$. A similar argument applies to $\chi_2(4y/x^2)M_1 \cdots M_j f(u)$ for $M_1, \ldots, M_j \in \mathcal{M}_2$. Q.E.D.

A commutator argument using a decomposition similar to that given in the proof of Theorem 4.8 now yields the propagation of regularity for waves which are of strongly conormal with respect to a tangent pair of characteristic hypersurfaces. See Figure 4.6. The natural time variable for the operator given by (4.9) is $t = y + z$. In order to avoid dealing with the initial value problem, we assume that the problem is linear in the far past.

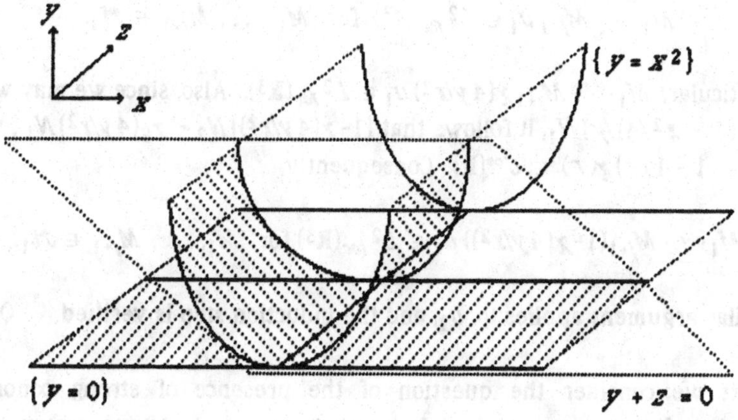

Figure 4.6

Theorem 4.16. *Suppose that* $u \in L^\infty_{loc}(\mathbf{R}^3)$, f *is smooth*, $f = 0$ *for* $y + z \ll 0$, *and* $p_2(D)u = (\partial_y\partial_z - \partial_x{}^2 + y\partial_y{}^2)u = f(x,y,z,u)$. *If* u *is strongly conormal with respect to the tangent pair* $\{y = 0\}, \{y = x^2/4\}$ *for* $y + z < 0$, *then* u *is strongly conormal with respect to* $\{y = 0\}, \{y = x^2/4\}$ *for all* (x,y,z).

Proof. The commutators of the operator with the vector fields given by (4.10) are

$$[p_2(D),M_0] = 2p_2(D),\ [p_2(D),M_1] = \partial_x(M_0 + M_2),\ [p_2(D),M_2] = 0,$$
$$[p_2(D),N_1] = \partial_y(M_3 + N_1 - 1),\ [p_2(D),N_2] = \partial_y(M_0 + M_3 - N_2).$$

Let χ be as in Definition 4.14, let u_1 be the solution to

$$p_2(D)u_1 = \chi(4y/x^2)f(x,y,z,u),\quad u_1 = 0 \text{ for } y + z \ll 0,$$

and set $u_2 = u - u_1$, so that $p_2(D)u_2 = (1 - \chi(4y/x^2))f(x,y,z,u)$.

Assume by induction on j that

$$M_1 \cdots M_j\chi(4y/x^2)u \in L^2{}_{loc}(\mathbf{R}^3) \text{ for } M_1, \ldots, M_j \in \mathcal{M}_1, \text{ and}$$
$$M_1 \cdots M_j(1-\chi(4y/x^2))u \in L^2{}_{loc}(\mathbf{R}^3) \text{ for } M_1, \ldots, M_j \in \mathcal{M}_2.$$

Then, from Lemma 4.15 and the usual commutator argument applied to the vector fields $\{M_0, M_1, M_2, N_1\}$,

$$M_1 \cdots M_{j+1} u_1 \in L^2{}_{loc}(\mathbf{R}^3) \text{ for } M_1, \ldots, M_{j+1} \in \mathcal{M}_1.$$

In particular, $M_1 \cdots M_{j+1}\chi(4y/x^2)u_1 \in L^2{}_{loc}(\mathbf{R}^3)$. Also, since we may write $N_2 = \{((y - x^2/4)/y)N_1$, it follows that $(1-\chi(4y/x^2))N_2 = \chi_2(4y/x^2)N_1$, with $\chi_2(r) = (1 - 1/r)\chi(r) \in C^\infty(\mathbf{R})$. Consequently,

$$M_1 \cdots M_{j+1}(1-\chi(4y/x^2))u_1 \in L^2{}_{loc}(\mathbf{R}^3) \text{ for } M_1, \ldots, M_{j+1} \in \mathcal{M}_1.$$

A similar argument applies to u_2, and the induction step is verified. Q.E.D.

Next we consider the question of the presence of strong conormal regularity after a transversal triple interaction, and employ an argument of Beals-Métivier which is similar to that used in the proof of Theorem 4.8. The definition of u being conormal in the past with respect to characteristic hyperplanes is taken to be the analogue of Definition 4.14; that is, u remains in $L^2{}_{loc}(\mathbf{R}^3)$ when differentiated by any vector fields tangent to the hyperplanes.

Theorem 4.17. *Let f be smooth, and suppose that $u \in L^\infty{}_{loc}(\mathbf{R}^3)$ satisfies $\Box u = f(x,y,z,u)$. Assume that u is conormal in the past with respect to three characteristic hyperplanes for $\Box$ which intersect transversally at the origin. Let N be a neighborhood of a point in the intersection of Σ_0, the surface of the light cone over the origin, with one of the hyperplanes Σ_1, such that N does not intersect the other hyperplanes. Then u is strongly conormal with respect to the tangent pair Σ_0, Σ_1 on N.*

Proof. Since we are now interested in the behavior of u near the intersection of only one of the hyperplanes and the surface of the light cone, we choose coordinates so that the $\Sigma_1 = \{y = 0\}$ and $\Sigma_0 = \{\alpha = 4yz - x^2 = 0\}$, with intersection $\{x = y = 0\}$, and the wave operator is $\Box = \partial_y\partial_z - \partial_x{}^2$. In these coordinates, the vector fields from Definition 4.9 which are tangent to $\{x = 0\}$ and $\{\alpha = 0\}$ may be written as

$$(4.11) \qquad M_0 = x\partial_x + y\partial_y + z\partial_z,\ M'_1 = x\partial_x + 2y\partial_y, \text{ and } M'_2 = x\partial_z + 2y\partial_x.$$

Let $\chi \in C^\infty(\mathbf{R}^3\backslash 0)$ be homogeneous of degree zero and have conic support near $\{y = 0\} \cap \{\alpha = 0\}$, away from $\{z = 0\}$, and let u be as in Theorem 4.17. The proof of Theorem 4.8 may be easily adapted to the conormal spaces in this context (which involve $L^2_{loc}(\mathbf{R}^3)$ rather than $H^{3/2-}$). In particular, it follows that

$$M_0^{\beta_0} M'_1{}^{\beta_1} M'_2{}^{\beta_2} \chi u \in L^2_{loc}(\mathbf{R}^3) \text{ for } |\beta| \le k. \tag{4.12}$$

In order to establish the stronger conormal regularity, we introduce a further localization, tailored to separating the tangential hypersurfaces but respecting the natural three dimensional homogeneity. Let $\chi'(\theta_1, \theta_2)$ be homogeneous of degree zero and smooth away from the origin, with

$$\chi'(\theta_1, \theta_2) = 0 \text{ on } \{\theta_2 \ge \theta_1/8\},$$
$$\chi'(\theta_1, \theta_2) = 1 \text{ on } \{\theta_2 \le \theta_1/16\} \cap \{\theta_2 \le \theta_1/4\}.$$

Then $\chi'(x^2-4yz, yz)$ is homogeneous of degree zero and smooth away from $\{x = y = 0\}$, is supported where $yz \le x^2/12$, and is identically one where $yz \le x^2/8$. See Figure 4.7, which shows the configuration for z fixed. Notice that $x^2 \le (x^2-4yz) + x^2/3$ on $supp(\chi'(x^2-4yz, yz))$, and hence

$$x^2 \le (3/2)(x^2-4yz) \text{ on } supp(\chi'(x^2-4yz, yz)). \tag{4.13}$$

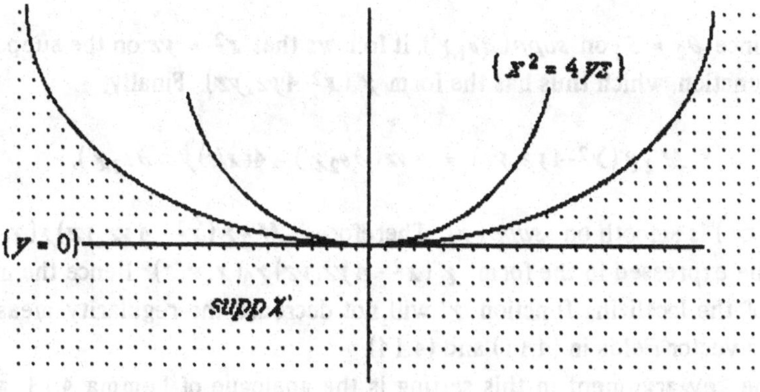

Figure 4.7

From the definition of strong conormality, it is clearly sufficient to prove that $\chi'(x^2-4yz, yz)\chi u$ is conormal with repect to the pair $\{y = 0\}, \{x = y = 0\}$, and that $(1 - \chi'(x^2-4yz, yz))\chi u$ is conormal with repect to the pair $\{\alpha = 0\}$, $\{x = y = 0\}$. We consider the first term, as the second is handled in the same fashion.

Again, we use vector fields with coefficients which are homogeneous of degree one and smooth on $supp(\chi)$ away from the origin. All such vector fields which are simultaneously tangent to $\{y = 0\}$ and $\{x = y = 0\}$ may be easily seen to be generated by $x\partial_x$, $y\partial_x$, $y\partial_y$, and $z\partial_z$, because $z \neq 0$ on $supp(\chi)$. Moreover, if regularity with respect to

$$N'_3 = x\partial_x \tag{4.14}$$

can be established, then the proof is finished, because (4.12) holds,

$$y\partial_y = (M'_1 - N'_3)/2,\ z\partial_z = M_0 - (M'_1 + N'_3)/2,\ y\partial_x = (M'_2 - (x/z)z\partial_z), \tag{4.15}$$

and (x/z) is smooth on $supp(\chi)$ away from the origin. Since

$$M_0\chi'(x^2-4yz,yz) = M'_1\chi'(x^2-4yz,yz) = 2((x^2-4yz)(\partial_{\theta_1}\chi') + yz(\partial_{\theta_2}\chi')),$$

these functions are again of the form $\chi''(x^2-4yz,yz)$, with $\chi''(\theta_1,\theta_2)$ homogeneous of degree zero and smooth away from the origin. Furthermore,

$$N'_3\chi'(x^2-4yz,yz) = 2x^2(\partial_{\theta_1}\chi') = 2(x^2/yz)yz(\partial_{\theta_1}\chi'),$$

and since $\theta_2 \approx \theta_1$ on $supp(\partial_{\theta_1}\chi')$, it follows that $x^2 \approx yz$ on the support of this function, which thus has the form $\chi''(x^2-4yz,yz)$. Finally,

$$M'_2\chi'(x^2-4yz,yz) = 4yz(\partial_{\theta_2}\chi') = 4(x/z)yz(\partial_{\theta_2}\chi'),$$

and (x/z) is smooth on $supp(\chi)$. Therefore, $M'_2\chi'(x^2-4yz,yz)\chi(x,y,z)$ may be expressed in the form $\chi''(x^2-4yz,yz)\chi_0(x,y,z)$. Hence the insertion of the localizing function χ' will not decrease the regularity measured by the vector fields in (4.11) and (4.14).

The key argument in this setting is the analogue of Lemma 4.13, which allows regularity with respect to the remaining vector field N'_3 to be established. It is proved by using the expressions for $y\partial_y$ and $z\partial_z$ given in (4.15) and by writing $4yz(\partial_y\partial_z - \partial_x{}^2)$ in terms of these operators and N'_3.

Lemma 4.18. *Modulo terms of first order in* M_0, M'_1, and N'_3,

$$(N'_3)^2 = (x^2/(x^2-4yz))(4yz\Box + (M'_1)^2 - 2M_0M'_1 + M_0N'_3).$$

From (4.13), the coefficient on the right in this expression is bounded on $supp(\chi'(x^2 - 4yz, yz))$. Just as in the proof of (4.8), if f is smooth and u satisfies $\Box u = f(x, y, z, u)$, the regularity of $\chi'(x^2 - 4yz, yz)\chi u$ may be deduced from (4.15) and Lemma 4.18. The proof of Theorem 4.17 is thus completed. Q.E.D.

The nonflat case is handled similarly, using suitable changes of coordinates. For a different approach, and the analysis of subsequent interactions involving solutions strongly conormal with respect to tangent pairs and additional characteristic hypersurfaces, see Sà Barreto [65].

Chapter V. Regularity and Singularities in Problems on Domains With Boundary

If $\Omega \subset \mathbf{R}^n$ is an open domain with smooth boundary, one considers the regularity of solutions to strictly hyperbolic equations of the form

$$p(t,x,D)u = f(t,x) \text{ on } \mathbf{R} \times \Omega$$

in terms of the regularity of f, the restriction of u to a spacelike initial surface or to a region in the past which determines the values of u in the future, and the restriction of u to the boundary $\mathbf{R} \times \partial\Omega$. A fundamental example is the following. Let Ω be the complement of the closed unit ball in $\mathbf{R}^n$, and solve

$$\begin{gathered} \Box u = 0 \text{ on } \mathbf{R} \times \Omega, \\ u|_{t=0} = g_0, \ \partial_t u|_{t=0} = g_1 \text{ on } \Omega, \\ u|_{\mathbf{R} \times \partial\Omega} = 0. \end{gathered} \tag{5.1}$$

The analogue of the linear energy inequality again holds in this setting: if

$$E(t) = \left(\int_\Omega |\partial_t u|^2 + |\nabla_x u|^2 \, dx \right)^{1/2},$$

integration by parts implies that $E(t)$ is constant if u is a suitably smooth solution of (5.1). This property may be used to derive the existence of solutions to (5.1), if for instance $g_0 \in H^1(\Omega)$, $g_1 \in H^0(\Omega)$. The existence results for a general strictly hyperbolic operator with appropriate conditions on the boundary values and on the initial data are given, for example, in Taylor [70] and Chazarain-Piriou [21]. Once again, finite propagation speed and smooth changes of local coordinates allow such problems to be reduced locally to the flat case, in which $\Omega = \mathbf{R}^n_+ = \{(x_1,x'): \ x_1 > 0\}$.

It is natural to consider the microlocal propagation of singularities in such a setting. If in (5.1) the initial data are chosen to have small compact support, then finite propagation speed implies that the solution will agree with the free space solution for small time. The simplest microlocal problem is one for which the free space solution has wavefront set consisting of a single null bicharacteristic Γ_1 for $\Box$. If the (t, x) projection of Γ_1 intersects $\mathbf{R} \times \partial\Omega$ transversally, it will be demonstrated below that the wavefront set of the solution to (5.1) is contained in the union of the original and the "reflected" null bicharacteristic. In particular, the singular support is as pictured in Figure 5.1. For the proof in the case of an operator of order m, see Nirenberg [54]. In order to describe these results, we will first analyze reflected families of null bicharacteristics, and then describe local and microlocal Sobolev spaces appropriate to problems on domains with boundary.

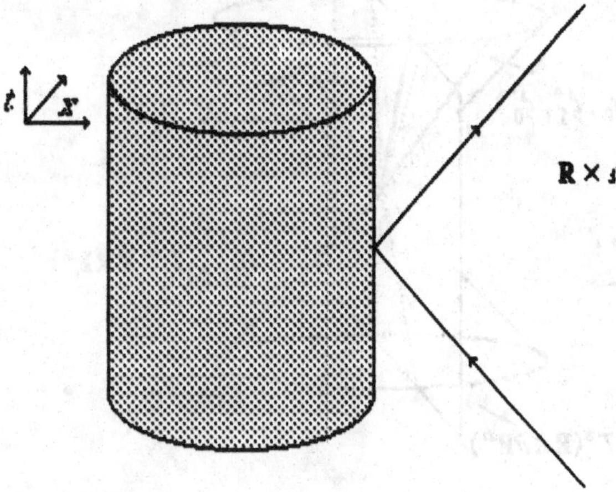

Figure 5.1

To compute a reflected null bicharacteristic in the case of the problem (5.1), it is simplest to change coordinates to flatten the boundary. Locally, the boundary will be given by $\{x_1 = 0\}$, and the intersection of the (t, x) projection of Γ_1 with the boundary may be taken to occur at the point $t = 0$, $x' = 0$. We can assume that the symbol $p_2(t, x, \tau, \xi)$ in the new coordinates satisfies $p_2(0,0,\tau,\xi) = \tau^2 - |x|^2$. If the null bicharacteristic is given by

$$(t(s), x(s), \tau(s), \xi(s)),\ t(0) = 0,\ x(0) = 0,$$

set $\xi_1(0) = \xi_1^-$. The cotangent space to $\mathbf{R} \times \partial\mathbf{R}^n_+$, considered as a subset of

$T^*(\mathbf{R}^{n+1})$, is given by $\{(t,0,x',\tau,0,\xi')\}$, and the projection of Γ_1 into this space is $(0,0,0,\tau(0),0,\xi'(0))$. There are either one or two distinct roots to the equation $[\tau(0)]^2 - ([\xi_1]^2 + [\xi'(0)]^2) = 0$; a single solution occurs if and only if $(\partial p_2/\partial\xi_1)(0,0,0,\tau(0),\xi_1^-,\xi'(0)) = 0$. Thus the existence of a distinct second root ξ_1^+ is equivalent to $(dx_1/ds)(0) \neq 0$, that is, the transversal intersection of the projection $\{(t(s),x(s))\}$ with $\mathbf{R}\times\partial\mathbf{R}^n_+$, and in this case, $\xi_1^- = -\xi_1^+$. There is a unique null bicharacteristic through $(0,0,0,\tau(0),\xi_1^+,\xi'(0))$; its restriction to $T^*(\mathbf{R}\times\mathbf{R}^n_+)$ is called the reflected null bicharacteristic. This procedure of projecting a subset of *char*(p_2) into $T^*(\mathbf{R}\times\partial\mathbf{R}^n_+)$ and then lifting the projection back to *char*(p_2) is illustrated in Figure 5.2, where the (τ,ξ) values over the origin are shown.

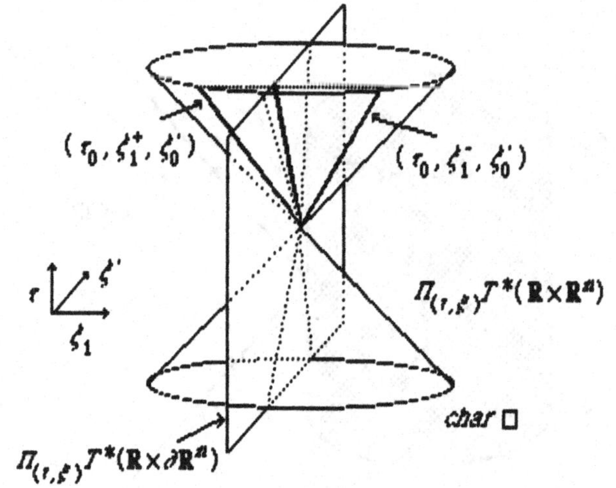

Figure 5.2

When $p_2(0,0,0,\tau(0),\xi_1^-,\xi'(0)) = 0$ has only a single solution, the null bicharacteristic projection $\{(t(s),x(s))\}$ is simply tangent to $\mathbf{R}\times\partial\mathbf{R}^n_+$ (because Ω in the original coordinates is the complement of a strictly convex set). In this instance the null bicharacteristic is said to be grazing, and the microlocal analysis is considerably more difficult. It was established by Taylor and by Melrose that microlocal regularity propagates along grazing null bicharacteristics; see Hörmander [37], for example.

For a higher order operator, a similar analysis may be performed. It is assumed that $\mathbf{R}\times\partial\Omega$ is non-characteristic for the operator. In this case, in the coordinates with flattened boundary given locally by $\{x_1 = 0\}$, let the strictly hyperbolic symbol be denoted by $p_m(t,x,\tau,\xi)$. Let a null bicharacteristic be given by $(t(s),x(s),\tau(s),\xi(s))$, with $t(0)=0$, $x(0)=0$, $\xi_1(0)=\xi_{1,1}$. There are k roots $\xi_{1,1},\dots,\xi_{1,k}$ of the equation $p_m(0,0,0,\tau(0),\xi_1,\xi'(0))=0$, with $1\le k\le m$. All of the roots are simple if and only if $(\partial p_m/\partial\xi_1)\neq 0$ at each of the points $(0,0,0,\tau(0),\xi_{1,j},\xi'(0))$, that is, the correponding null bicharacteristics $\Gamma_1,\dots,\Gamma_k$ all have (t,x) projections which intersect $\mathbf{R}\times\partial\mathbf{R}^n_+$ transversally. See Figure 5.3 for the (τ,ξ) projections of two examples for fourth order equations. The corresponding reflected families have cardinalities two and four.

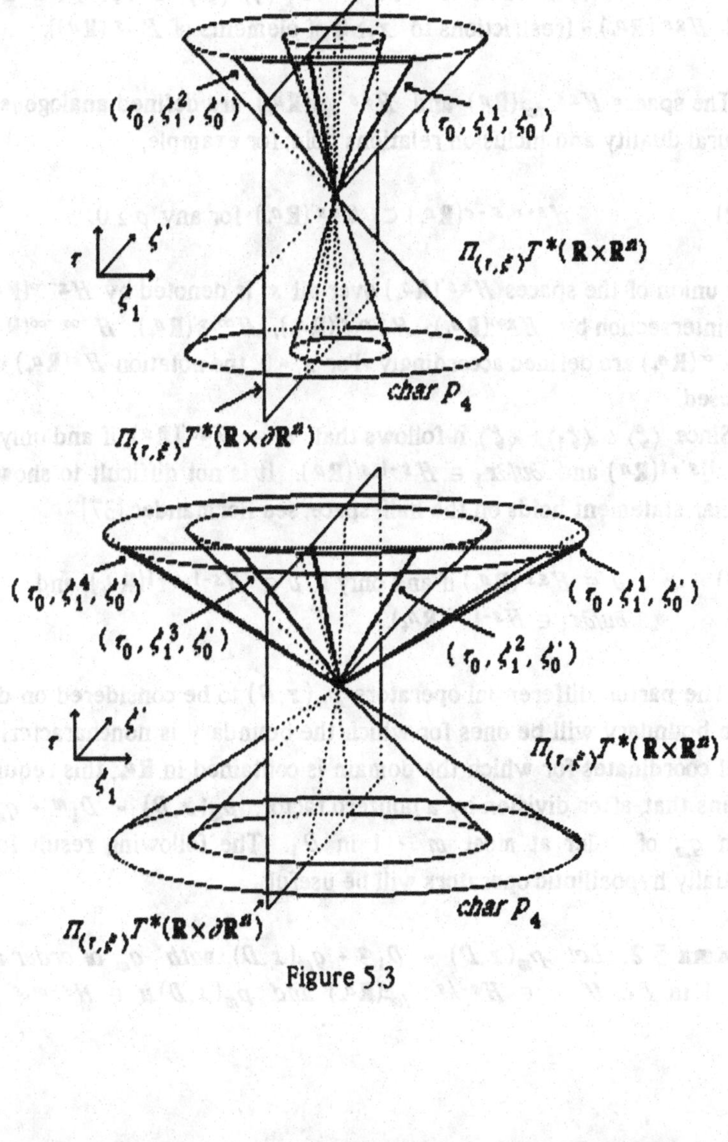

Figure 5.3

In order to describe precisely the propagation of regularity in a region with boundary, appropriate notions of local and microlocal Sobolev spaces are necessary. The natural local definition involves Hörmander's spaces of functions of mixed Sobolev regularity with repect to the directions tangent and normal to the boundary. The simplest microlocal definition involves tangential pseudodifferential operators. The regularity involved in these spaces is coordinate dependent in general, but for solutions of appropriate partial differential equations it is not.

Definition 5.1. Let the coordinates in $\mathbf{R}^n$ be denoted by (x_1,x'), with dual coordinates (ξ_1,ξ'). Then $H^{s,s'}(\mathbf{R}^n) = \{u: \langle\xi\rangle^s\langle\xi'\rangle^{s'}u^\wedge(\xi_1,\xi') \in L^2(\mathbf{R}^n)\}$, and $H^{s,s'}(\mathbf{R}^n_+) = \{$restrictions to $x_1 \geq 0$ of elements of $H^{s,s'}(\mathbf{R}^n)\}$.

The spaces $H^{s,s'}{}_{loc}(\mathbf{R}^n)$ and $H^{s,s'}{}_{loc}(\mathbf{R}^n_+)$ are defined analogously. The natural duality and inclusion relations hold; for example,

$$H^{s+\sigma,s'-\sigma}(\mathbf{R}^n_+) \subset H^{s,s'}(\mathbf{R}^n_+) \text{ for any } \sigma \geq 0. \tag{5.2}$$

The union of the spaces $H^{s,s'}(\mathbf{R}^n_+)$ over all s' is denoted by $H^{s,-\infty}(\mathbf{R}^n_+)$, and the intersection by $H^{s,\infty}(\mathbf{R}^n_+)$; $H^{-\infty,s'}(\mathbf{R}^n_+)$, $H^{\infty,s'}(\mathbf{R}^n_+)$, $H^{-\infty,-\infty}(\mathbf{R}^n_+)$, and $H^{\infty,-\infty}(\mathbf{R}^n_+)$ are defined accordingly. For $s' = 0$, the notation $H^s(\mathbf{R}^n_+)$ will also be used.

Since $\langle\xi\rangle \leq \langle\xi_1\rangle + \langle\xi'\rangle$, it follows that $u \in H^{s,s'}(\mathbf{R}^n)$ if and only if $u \in H^{s-1,s'+1}(\mathbf{R}^n)$ and $\partial u/\partial x_1 \in H^{s-1,s'}(\mathbf{R}^n)$. It is not difficult to show that a similar statement holds on the half-space; see Hörmander [37].

(5.3) $u \in H^{s,s'}(\mathbf{R}^n_+)$ if and only if $u \in H^{s-1,s'+1}(\mathbf{R}^n_+)$ and $\partial u/\partial x_1 \in H^{s-1,s'}(\mathbf{R}^n_+)$.

The partial differential operators $p_m(x,D)$ to be considered on domains with boundary will be ones for which the boundary is noncharacteristic. In local coordinates for which the domain is contained in $\mathbf{R}^n_+$, this requirement means that, after division by a nonzero factor, $p_m(x,D) = D_1{}^m + q_m(x,D)$, with q_m of order at most $m - 1$ in D_1. The following result for such partially hypoelliptic operators will be useful.

Lemma 5.2. *Let* $p_m(x,D) = D_1{}^m + q_m(x,D)$, *with* q_m *of order at most* $m - 1$ *in* D_1. *If* $u \in H^{s-j,s'+j}{}_{loc}(\mathbf{R}^n_+)$ *and* $p_m(x,D)u \in H^{s-m,s'}{}_{loc}(\mathbf{R}^n_+)$, *then* $u \in H^{s,s'}{}_{loc}(\mathbf{R}^n_+)$.

Proof. By (5.2), it may be assumed that j is an integer, $j \geq 1$. From the assumption on q_m and (5.2), we have $q_m(x,D)u \in H^{s-j-m+1,s'+j-1}{}_{loc}(\mathbf{R}^n_+)$. Hence the assumptions on p_m, j, and $p_m(x,D)u$ imply that

$$(D_1{}^m)u \in H^{s-j-m+1,s'+j-1}{}_{loc}(\mathbf{R}^n_+).$$

Since $(D_1{}^{m-1})u \in H^{s-j-m+1,s'+j}{}_{loc}(\mathbf{R}^n_+)$, it follows from (5.3) that

$$(D_1{}^{m-1})u \in H^{s-j-m+2,s'+j-1}{}_{loc}(\mathbf{R}^n_+).$$

By induction on k, it is a consequence of $(D_1{}^{m-k})u \in H^{s-j-m+k,s'+j}{}_{loc}(\mathbf{R}^n_+)$ and (5.3) that

$$(D_1{}^{m-k})u \in H^{s-j-m+k+1,s'+j-1}{}_{loc}(\mathbf{R}^n_+).$$

Therefore, $u \in H^{s-j+1,s'+j-1}{}_{loc}(\mathbf{R}^n_+)$, and the result now follows by induction on j. Q.E.D.

Microlocal definitions are made globally on $\mathbf{R}^n_+$ in terms of boundary pseudodifferential operators of the form $a(x_1,x',D_{x'})$, with symbols which are classical in the variables (x',ξ') and depend smoothly on x_1. In the corresponding pseudodifferential calculus, the remainder terms, with symbols $a(x_1,x',\xi') \in S^{-\infty}{}_{1,0}$, will in general be smoothing in x' only. In particular, they map $H^{s,-\infty}(\mathbf{R}^n_+)$ into $H^{s,\infty}(\mathbf{R}^n_+)$.

Definition 5.3. Let $(0,x'_0,\xi'_0) \in T^*(\partial\mathbf{R}^n_+)$ and suppose that $u \in H^{s,-\infty}(\mathbf{R}^n_+)$. Then $u \in H^{s,s'}{}_{ml}(0,x'_0,\xi'_0)$ if there is a boundary pseudodifferential operator $a(x_1,x',D_{x'})$ of order zero, microlocally elliptic at $(0,x'_0,\xi'_0)$, such that $a(x_1,x',D_{x'})u \in H^{s,s'}(\mathbf{R}^n_+)$.

The wavefront set up to the boundary for distributions in $H^{\infty,-\infty}(\mathbf{R}^n_+)$ is defined in the natural fashion, extending the notion on the interior.

Definition 5.4. For a distribution $u \in H^{\infty,-\infty}(\mathbf{R}^n_+)$, the wavefront set is the subset of $T^*(\mathbf{R}^n_+) \cup T^*(\partial\mathbf{R}^n_+)$ determined as follows: in $T^*(\mathbf{R}^n_+)$, $WF(u)$ is the set given by Definition 1.1, while in $T^*(\partial\mathbf{R}^n_+)$, $(0,x'_0,\xi'_0) \notin WF(u)$ if

$u \in H^{\infty}{}_{ml}(0, x'_0, \xi'_0)$.

The precise regularity of a solution to (5.1) which is assumed to have wave front set in the past along a single non-grazing null bicharacteristic may now be described.

Proposition 5.5. *Let Ω be the complement of the unit ball in $\mathbf{R}^n$. Suppose that $u \in H^s(\mathbf{R}\times\Omega)$ satisfies $\Box u = 0$ on $\mathbf{R}\times\Omega$, $u|_{\mathbf{R}\times\partial\Omega} = 0$. Let Γ_1 be a null bicharacteristic for $\Box$ with (t,x) projection which intersects $\mathbf{R}\times\partial\Omega$ transversally at $(0, x_0)$, let γ be the projection of Γ_1 onto $T^*(\mathbf{R}\times\partial\Omega)$, and let Γ_2 be the reflected null bicharacteristic. Assume that, for some $t_0 < 0$, $WF(u)\cap\{t < t_0\} \subset \Gamma_1\cap\{t < t_0\}$. Then $WF(u)\cap T^*(\mathbf{R}\times\Omega) \subset \{\Gamma_1\cup\Gamma_2\}\cap T^*(\mathbf{R}\times\Omega)$, and $WF(u)\cap T^*(\mathbf{R}\times\partial\Omega) \subset \gamma$.*

Proof. By finite propagation speed, it suffices to work on a small neighborhood of $(0, x_0)$. By Theorem 1.8, if u' is a solution to $\Box u' = 0$ on $\mathbf{R}\times\mathbf{R}^n$ which agrees with u on a neighborhood in $\{t < t_0\}$ of the (t,x) projection of Γ_1, then $WF(u') \subset \Gamma_1$. The idea is to show that u'' with the following properties may be constructed:

$$\Box u'' = 0 \text{ on } \mathbf{R}\times\mathbf{R}^n,\ WF(u'') \subset \Gamma_2,\ u''|_{\mathbf{R}\times\partial\Omega} = -u'|_{\mathbf{R}\times\partial\Omega} \text{ near } (0, x_0).$$

It will then follow that $u - (u' + u'')$ is smooth on a neighborhood of $(0, x_0)$ in $\mathbf{R}\times\Omega$, and the description of the wave front set will be complete.

In the flattened coordinates near $(0, x_0)$ described previously, we have $\Omega = \mathbf{R}^n_+$, $x_0 = (0,0) \in \partial\mathbf{R}^n_+$, and $\Box$ has symbol $p_2(t, x, \tau, \xi_1, \xi')$, with

$$p_2(0,0,\tau,\xi) = (\tau^2 - |\xi'|^2) - (\xi_1)^2.$$

Since $u \in H^s{}_{loc}(\mathbf{R}^{n+1}{}_+)$ and $p_2(t, x, D_t, D_x)\,u = 0$, Lemma 5.2 implies that $u \in H^{\infty,-\infty}{}_{loc}(\mathbf{R}\times\Omega)$. The null bicharacteristic Γ_1 passes through the point $(0,0,\tau_0, x_1^-, \xi'_0)$, with $\xi_1^- \neq 0$, and $\gamma = (0,0,\tau_0,\xi'_0)$. As in Nirenberg [54], the problem is most easily treated by reduction to a first order system. Set $z = (t, x')$, and denote the dual variable by ζ. Let $K \subset T^*(\mathbf{R}^{n+1}{}_+)$ be a conic neighborhood of $(0,0,0,\zeta_0)$, with (ξ_1, ζ) projection of the form $\{|\zeta - |\zeta_0| \le \varepsilon|\zeta_0|\}$. The given expression for p_2 implies that there are classical pseudodifferential symbols $\lambda_\pm(x_1, z, \zeta)$ of order one, with real principal part, such that

(5.4) $$p_2 = -(\xi_1 - \lambda_-(x_1,z,\zeta))(\xi_1 - \lambda_+(x_1,z,\zeta)) \text{ on } K,$$

and for some positive constant c,

(5.5) $$|\lambda_\pm(x_1,z,\zeta)| \geq c|\zeta| \text{ and } |\lambda_+(x_1,z,\zeta) - \lambda_-(x_1,z,\zeta)| \geq c|\zeta|.$$

Near the origin, Γ_1 is the null bicharacteristic for $\xi_1 - \lambda_+(x_1,z,\zeta)$ through $(0,0,\xi_1^-,\zeta_0)$, and Γ_2 is the null bicharacteristic for $\xi_1 - \lambda_-(x_1,z,\zeta)$ through $(0,0,\xi_1^+,\zeta_0)$. We will be interested in treating first order systems of the form $(D_1 - \lambda_\pm(x_1,z,D_z))\, w_\pm = f_\pm$.

Lemma 5.6. *Let $\lambda(x_1,z,D_z)$ be a classical properly supported pseudodifferential operator of order one with real principal symbol. Set $\gamma = (0,0,\zeta_0)$, and let Γ be the null bicharacteristic in $\mathbf{R}^{n+1}{}_+$ for $\xi_1 - \lambda(x_1,z,\zeta)$ through $(0,0,\lambda(0,0,\zeta_0),\zeta_0)$. If $f \in H^s{}_{loc}(\mathbf{R}^{n+1}{}_+)$, and $g \in H^s(\partial\mathbf{R}^{n+1})$ has compact support, then there is a unique solution $w \in H^s{}_{loc}(\mathbf{R}^{n+1}{}_+)$ to*

$$(D_1 - \lambda(x_1,z,D_z))w = f,\ u(0,z) = g.$$

If, additionally, $f \in H^{\infty,-\infty}(\mathbf{R}^{n+1}{}_+)$, $WF(f) \subset \gamma \cup \Gamma$, and $WF(g) \subset \gamma$, then $w \in H^{\infty,-\infty}(\mathbf{R}^{n+1}{}_+)$ and $WF(w) \subset \gamma \cup \Gamma$.

Proof. If $w \in C(\mathbf{R}_+;H^s(\mathbf{R}^n))$, define the energy $E(x_1)$ to be the norm of $w(x_1,z)$ in $H^s(\mathbf{R}^n)$, and let $F(x_1)$ to be the norm of $(D_1 - \lambda(x_1,z,D_z))\,w$ in $H^s(\mathbf{R}^n)$. If w has compact support in z for each x_1, integration by parts and the assumption that $\lambda(x_1,z,D_z) = \lambda^*(x_1,z,D_z)$ modulo terms of order zero imply that $dE/dx_1 \leq C(x_1)(E(x_1) + F(x_1))$. Therefore, by Gronwall's inequality,

$$E(x_1) \leq C_1(x_1)\Big\{E(0) + \int_0^{x_1} F(s)C_2(s)\,ds\Big\}.$$

It follows by a contraction mapping argument that, for f and g as in the hypotheses, there is a unique solution $w \in C(\mathbf{R}_+;H^s(\mathbf{R}^n))$ to

$$(D_1 - \lambda(x_1,z,D_z))u = f,\ \ u(0,z) = g.$$

Since $w \in H^{0,s}{}_{loc}(\mathbf{R}^{n+1}{}_+)$, Lemma 5.2 yields that $w \in H^{s+1,-1}{}_{loc}(\mathbf{R}^{n+1}{}_+)$, so in particular $w \in H^s{}_{loc}(\mathbf{R}^{n+1}{}_+)$. If $f \in H^{\infty,-\infty}(\mathbf{R}^{n+1}{}_+)$, the same argument implies that $w \in H^{\infty,-\infty}(\mathbf{R}^{n+1}_+)$.

If $WF(g) \subset \gamma$, let K be a small conic neighborhood of $(0,0,\lambda(0,0,\zeta_0),\zeta_0)$ in $T^*(\mathbf{R}^{n+1}_+)$ with (ξ_1,ζ) projection of the form $\{|\zeta - \zeta_0| \le \epsilon|\zeta_0|\}$, and let U be a small neighborhood of 0 in $\mathbf{R}^{n+1}_+$. Let $e(x_1,z,\zeta)$ be a pseudodifferential cutoff function which is of order zero and vanishes on K. A commutator argument and the energy estimate above easily yield that $e(x_1,z,D_z)w \in H^\infty_{loc}(U)$. Therefore, $WF(w) \cap T^*(\partial\mathbf{R}^{n+1}_+) \subset \gamma$ and $WF(w) \cap T^*U \subset K$. Since $\xi_1 - \lambda(x_1,z,\zeta)$ is microlocally elliptic on K except on a small neighborhood of Γ, it follows that $WF(w) \cap T^*U \subset \Gamma$. The interior regularity result (Theorem 1.8) then implies that $WF(w) \cap T^*(\mathbf{R}^{n+1}_+) \subset \Gamma$. Q.E.D.

Lemma 5.7. *Let $\lambda(x_1,z,D_z)$, γ, and Γ be as in Lemma 5.6. Suppose that $w \in H^s_{loc}(\mathbf{R}^{n+1}_+)$ and $(D_1 - \lambda(x_1,z,D_z))w \in H^s_{loc}(\mathbf{R}^{n+1}_+)$. Then $w(0,z) \in H^s_{loc}(\partial\mathbf{R}^{n+1})$. If, additionally, $(D_1 - \lambda(x_1,z,D_z))w \in H^{\infty,-\infty}(\mathbf{R}^{n+1}_+)$ and $WF(w) \cap T^*(\mathbf{R}^{n+1}_+) \subset \Gamma$, then $WF(w(0,z)) \subset \gamma$.*

Proof. Since $w \in H^{0,s}_{loc}(\mathbf{R}^{n+1}_+)$, it follows that $w(x_0,z) \in H^s_{loc}(\mathbf{R}^n)$ for almost every x_0. Let $f \in H^s_{loc}(\mathbf{R}^{n+1})$ be an extension of $(D_1 - \lambda(x_1,z,D_z))w$ and let $\lambda(x_1,z,D_z)$ be extended to all of $\mathbf{R}^{n+1}$, still satisfying the hypotheses of Lemma 5.6. If w' is the solution to

$$(D_1 - \lambda(x_1,z,D_z))w' = f \text{ on } \{(x_1,z)\colon x_1 < x_0\},\ w'(x_0,z) = w(x_0,z),$$

then by Lemma 5.6, $w'(0,z) = w(0,z)$ and $w'(0,z) \in H^s_{loc}(\partial\mathbf{R}^{n+1})$. The statement about wave front sets may be deduced in the same fashion. Q.E.D.

In order to complete the proof of Proposition 5.5, we let p_2 be an extension of the operator in flattened coordinates to all of $\mathbf{R}^{n+1}$ and let u' be a solution in the flattened coordinates to $p_2 u' = 0$ on $\mathbf{R}^{n+1}$ which agrees with u on a neighborhood of a point in the projection of Γ_1 in $\{x_1 > 0,\ t < t_0\}$. With $\lambda_\pm$ as in (5.4) extended to all of $\mathbf{R}^{n+1}$ to satisfy (5.5), set

$$U = (u_1, u_2) = (\lambda_+(x_1,z,D_z)u', D_1 u').$$

Then $U \in H^{s-1}_{loc}(\mathbf{R}^{n+1}) \cap H^{\infty,-\infty}_{loc}(\mathbf{R}^{n+1})$ satisfies

$$(D_1 - A(x_1,z,D_z))U = 0,$$

with A a first order system of pseudodifferential operators. Moreover, by (5.4), it follows that, on K, $A(x_1,z,D_z)$ has principal symbol

$$A_1(x_1,z,\zeta) = \begin{bmatrix} \xi_1 & -\lambda_+ \\ \lambda_- & \xi_1-(\lambda_-+\lambda_+) \end{bmatrix}.$$

The calculus of pseudodifferential operators may be employed in the usual fashion to construct an invertible operator $Q(x_1,z,D_z)$ of order zero, with principal symbol

$$Q_0(x_1,z,\zeta) = \begin{bmatrix} \dfrac{\lambda_+}{\lambda_+-\lambda_-} & 1 \\ \dfrac{\lambda_-}{\lambda_+-\lambda_-} & 1 \end{bmatrix},$$

such that, modulo a term mapping $H^{\infty,-\infty}(\mathbf{R}^{n+1}{}_+)$ into $H^{\infty}(\mathbf{R}^{n+1}{}_+)$,

$$(5.6)\qquad Q^{-1}(D_1-A)Q = D_1-\Lambda \ \text{ on } K,\ \text{with } \Lambda = \begin{bmatrix} \lambda_+ & 0 \\ 0 & \lambda_- \end{bmatrix}.$$

Let $W' = Q(x_1,z,D_z)U'$. Then $W' \in H^{s-1}{}_{loc}(\mathbf{R}^{n+1}) \cap H^{\infty,-\infty}{}_{loc}(\mathbf{R}^{n+1})$, $WF(W') \subset \Gamma_1$, and from (5.6), $(D_1-\Lambda)W' \in H^{\infty}(\mathbf{R}^{n+1})$. From Lemma 5.7, $W'(0,z) \in H^{s-1}{}_{loc}(\partial\mathbf{R}^{n+1}{}_+)$, and $WF(W'(0,z)) \subset \gamma$.

If W'' is a solution to

$$(D_1-\Lambda)W'' = 0 \ \text{ on } \mathbf{R}^{n+1}{}_+,\ W''(0,z) = (0,w_2(z)),$$
$$w_2(z) \in H^{s-1}{}_{loc}(\mathbf{R}^{n}),\ WF(w_2(z)) \subset \gamma,$$

then by Lemma 5.6,

$$W'' \in H^{s-1}{}_{loc}(\mathbf{R}^{n+1}{}_+) \cap H^{\infty,-\infty}{}_{loc}(\mathbf{R}^{n+1}{}_+),\ WF(W'') \subset \Gamma_2.$$

For $U'' = (v_1,v_2) = Q^{-1}(x_1,z,D_z)W''$, we would like to be able to take boundary value $v_1(0,z) = -u_1(0,z)$. As a consequence of the expression for

Q, the equation $(0, w_2(z)) = Q(-u_1(0,z), v_2)$ can be solved. Modulo lower order terms,

$$v_2 = -\{\lambda_+/(\lambda_+ - \lambda_-)\} u_1(0,z),$$

and then

$$w_2 = -\{(\lambda_- + \lambda_+)/(\lambda_+ - \lambda_-)\} u_1(0,z).$$

If we let $u'' = \lambda_+(x_1, z, D_z)^{-1} v_1$, then $u'' \in H^s{}_{loc}(\mathbf{R}^{n+1}) \cap H^{\infty,-\infty}{}_{loc}(\mathbf{R}^{n+1})$, $WF(u'') \subset \Gamma_2$, $p_2 u'' \in H^\infty(\mathbf{R}^{n+1}{}_+)$, and $u'(0,z) + u''(0,z) \in H^\infty(\partial\mathbf{R}^{n+1}{}_+)$. Since $u' + u'' - u$ is smooth for $t < t_0$, and is smooth on $\partial\mathbf{R}^{n+1}{}_+$, we have $u' + u'' - u \in H^\infty(\mathbf{R}^{n+1}{}_+)$. Therefore,

$$WF(u) \cap T^*(\mathbf{R} \times \Omega) \subset WF(u') \cup WF(u'') \subset (\Gamma_1 \cup \Gamma_2) \cap T^*(\mathbf{R} \times \Omega),$$

and

$$WF(u) \cap T^*(\mathbf{R} \times \partial\Omega) \subset WF(u'(0,z)) \cup WF(u''(0,z)) \subset \gamma.$$

Proposition 5.5 is thereby established. Q.E.D.

The analogue of Hörmander's Theorem on the propagation of regularity along non-grazing null bicharacteristics for a strictly hyperbolic operator of order m may be derived in a similar fashion. In the notation used previously, it may be assumed that, locally near $(t = 0,\ x = 0)$, $\Gamma_1, \ldots, \Gamma_j$ have x_1 decreasing as t increases, while $\Gamma_{j+1}, \ldots, \Gamma_k$ have x_1 increasing as t increases. (Strict hyperbolicity with respect to t implies that t may be taken as the parameter on the null bicharacteristics.) If $k < m$, the corresponding first order system will include components which are solutions to forward and backward elliptic pseudodifferential equations, in addition to the hyperbolic pieces treated in Proposition 5.5. See, for example, Taylor [70] for the proof. As usual, the global hypotheses may be replaced with suitable local ones.

Theorem 5.8 (Lax-Nirenberg). *Let $p(t,x,D)$ be a strictly hyperbolic partial differential operator of order m on $\mathbf{R} \times \mathbf{R}^n_+$ such that $\mathbf{R} \times \partial\mathbf{R}^n_+$ is non-characteristic for p_m. Let Γ_1 be a null bicharacteristic for p_m which* *intersects* $\partial T^*(\mathbf{R}\times\mathbf{R}^n_+)$ *at* $(0,0,\tau_0,\xi_0)$. *Let* $\Gamma_1,\ldots,\Gamma_j,\Gamma_{j+1},\ldots,\Gamma_k$ *be the reflected family of null bicharacteristics, numbered as above, and assume that the* (t,x) *projections all intersect* $\mathbf{R}\times\partial\mathbf{R}^n_+$ *transversally. Let* $f(t,x)\in H^{s-m+1,-\infty}{}_{loc}(\mathbf{R}^n_+)\cap H^{s-m+1}{}_{ml}(0,0,\tau_0,\xi_0')\cap H^{s-m+1}{}_{ml}(\Gamma_1\cup\ldots\cup\Gamma_k)$. *Suppose that* u *is an extendible distribution on* $\mathbf{R}\times\mathbf{R}^n_+$ *satisfying*

$$p(t,x,D)u=f(t,x),$$
$$(\partial x_1)^i u\big|_{x_1=0}\in H^{s-i}{}_{ml}(0,0,\tau_0,\xi_0'),\ 0\le i\le[(m+k)/2]-j-1,\ \textit{and}$$
$$u\in H^s{}_{ml}(\Gamma_1\cup\ldots\cup\Gamma_j).$$

Then $u\in H^s{}_{ml}(\Gamma_{j+1}\cup\ldots\cup\Gamma_k)$.

If we wish to extend this result to nonlinear equations, the algebra properties of $H^{s,s'}(\mathbf{R}^n_+)$ and of $H^r{}_{ml}(x,\xi)$ in the boundary case are crucial. Schauder's Lemma in this setting is established in Sablé-Tougeron [67].

Lemma 5.9 (Sablé-Tougeron). *If* $u\in H^{s,s'}{}_{loc}(\mathbf{R}^n_+)$ *for* $s>1/2$, $s+s'>n/2$, $s+2s'>1/2$, *and if* $f(x,v)$ *is a* C^∞ *function of its arguments, then* $f(x,u)\in H^{s,s'}{}_{loc}(\mathbf{R}^n_+)$.

Proof. We show that if $u,v\in H^{s,s'}{}_{loc}(\mathbf{R}^n_+)$, then $uv\in H^{s,s'}{}_{loc}(\mathbf{R}^n_+)$. The general case may be derived using an argument like that given in the proof of Lemma 1.5. It may be assumed that u and v have been extended to be elements of $H^{s,s'}{}_{loc}(\mathbf{R}^n)$.

First, suppose that $s'\ge 0$. Then we may write

$$\langle\xi\rangle^s\langle\xi'\rangle^{s'}(uv)^\wedge(\xi)=\int K(\xi,\eta)f^\wedge(\xi-\eta)g^\wedge(\eta)\,d\eta,$$

with $f,g\in L^2(\mathbf{R}^n)$ and $K(\xi,\eta)=\langle\xi\rangle^s\langle\xi'\rangle^{s'}\langle\xi-\eta\rangle^{-s}\langle\xi'-\eta'\rangle^{-s'}\langle\eta\rangle^{-s}\langle\eta'\rangle^{-s'}$. We have

$$\begin{aligned}|K(\xi,\eta)|&\le C\langle\xi'\rangle^{s'}\{\langle\xi'-\eta'\rangle^{-s'}\langle\eta\rangle^{-s}\langle\eta'\rangle^{-s'}+\langle\xi-\eta\rangle^{-s}\langle\xi'-\eta'\rangle^{-s'}\langle\eta'\rangle^{-s'}\}\\&\le C\{\langle\eta\rangle^{-s}\langle\eta'\rangle^{-s'}+\langle\xi'-\eta'\rangle^{-s'}\langle\eta\rangle^{-s}+\langle\xi-\eta\rangle^{-s}\langle\eta'\rangle^{-s'}+\langle\xi-\eta\rangle^{-s}\langle\xi'-\eta'\rangle^{-s'}\}\\&\le C\langle\eta_1\rangle^{-1/2+\varepsilon}\{\langle\eta'\rangle^{-(s+s'-1/2+\varepsilon)}+\langle\xi'-\eta'\rangle^{-s'}\langle\eta'\rangle^{-(s-1/2+\varepsilon)}\}\\&\qquad+C\langle\xi_1-\eta_1\rangle^{-1/2+\varepsilon}\{\langle\xi'-\eta'\rangle^{-(s-1/2+\varepsilon)}\langle\eta'\rangle^{-s'}+\langle\xi'-\eta'\rangle^{-(s+s'-1/2+\varepsilon)}\}.\end{aligned}$$

If ε is sufficiently small, then $s+s'-1/2+\varepsilon>(n-1)/2$, and it follows from a simple modification of Lemma 1.4 that $\langle\xi\rangle^s\langle\xi'\rangle^{s'}(uv)^\wedge(\xi)\in L^2(\mathbf{R}^n)$.

If $s' < 0$, the corresponding kernel in the above integral has the form $K(\xi,\eta) = \langle\xi\rangle^{s}\langle\xi' - \eta'\rangle^{|s'|}\langle\eta'\rangle^{|s'|}\langle\xi - \eta\rangle^{-s}\langle\eta\rangle^{-s}\langle\xi'\rangle^{-|s'|}$. Then

$$\begin{aligned}|K(\xi,\eta)| &\leq C\langle\xi' - \eta'\rangle^{|s'|}\langle\eta'\rangle^{|s'|}\{\langle\eta\rangle^{-s}\langle\xi'\rangle^{-|s'|} + \langle\xi - \eta\rangle^{-s}\langle\xi'\rangle^{-|s'|}\}\\ &\leq C\langle\eta\rangle^{-s}\{\langle\eta'\rangle^{|s'|} + \langle\eta'\rangle^{2|s'|}\langle\xi'\rangle^{-|s'|}\} + C\langle\xi - \eta\rangle^{-s}\{\langle\eta'\rangle^{|s'|} + \langle\eta'\rangle^{2|s'|}\langle\xi'\rangle^{-|s'|}\}\\ &\leq C\langle\eta_1\rangle^{-1/2+\varepsilon}\{\langle\eta'\rangle^{-(s+s'-1/2+\varepsilon)} + \langle\eta'\rangle^{-(s+2s'-1/2+\varepsilon)}\langle\xi'\rangle^{-|s'|}\}\\ &\qquad + C\langle\xi_1 - \eta_1\rangle^{-1/2+\varepsilon}\{\langle\eta'\rangle^{-(s+s'-1/2+\varepsilon)} + \langle\eta'\rangle^{-(s+2s'-1/2+\varepsilon)}\langle\xi'\rangle^{-|s'|}\}.\end{aligned}$$

If ε is sufficiently small, then $s + s' - 1/2 + \varepsilon > (n-1)/2$ and $s + 2s' - 1/2 + \varepsilon > 0$. Again, from the adaptation of Lemma 1.4, $\langle\xi\rangle^{s}\langle\xi'\rangle^{s'}(uv)^{\wedge}(\xi) \in L^2(\mathbf{R}^n)$. Q.E.D.

Let $p(t,x,D)$ be a strictly hyperbolic partial differential operator of order m on $\mathbf{R} \times \mathbf{R}^n_+$ and let $u \in H^{s}_{loc}(\mathbf{R} \times \mathbf{R}^n_+)$, $s > (n+1)/2 + m - 2$, be a solution of the semilinear equation

$$p(t,x,D)u = f(t,x,u,\ldots,D^{m-2}u) \text{ on } \mathbf{R} \times \mathbf{R}^n_+. \tag{5.7}$$

As noted in David-Williams [30], Lemma 5.2 and Schauder's Lemma imply that $u \in H^{s+2,-2}_{loc}(\mathbf{R} \times \mathbf{R}^n_+)$, and by induction and Lemma 5.9 it follows that $u \in H^{2s-m+4-\rho,-s+m-4+\rho}_{loc}(\mathbf{R} \times \mathbf{R}^n_+)$ for any $\rho > 1/2$. This property suggests an appropriate space in which to prove the analogue of the microlocal algebra statement in Corollary 2.10. It may be proved using estimates similar to those of Lemma 1.9.

Lemma 5.10. *Let* $(x_0,\xi_0) \in T^*(\mathbf{R}^n_+) \cup T^*(\partial\mathbf{R}^n_+)$. *If* $1/2 < \sigma < s - n/2$, *then* $H^{2s-\sigma,-s+\sigma}_{loc}(\mathbf{R}^n_+) \cap H^{r}_{ml}(x_0,\xi_0)$ *is an algebra for* $(n+1)/2 < s \leq r < 2s - n/2 - \sigma$.

Just as Theorem 1.11 may be deduced from the microlocal algebra property and the linear regularity estimate, the microlocal propagation result for a simple semilinear problem on a domain with boundary may be obtained away from grazing null bicharacteristics using Theorem 5.8 and Lemma 5.10.

Theorem 5.11. *Let* $p(t,x,D)$ *be a strictly hyperbolic partial differential operator of order* m *on* $\mathbf{R} \times \mathbf{R}^n_+$ *such that* $\mathbf{R} \times \partial\mathbf{R}^n_+$ *is non-characteristic for* p_m. *Let* Γ_1 *be a null bicharacteristic for* p_m *which intersects* $\partial T^*(\mathbf{R} \times \mathbf{R}^n_+)$

at $(0,0,\tau_0,\xi_0)$. *Let* $\Gamma_1,\ldots,\Gamma_j,\Gamma_{j+1},\ldots,\Gamma_k$ *be the reflected family of null bicharacteristics, numbered as above, and assume that the* (t,x) *projections all intersect* $\mathbf{R}\times\partial\mathbf{R}^n_+$ *transversally. Suppose that* $u \in H^s_{loc}(\mathbf{R}\times\mathbf{R}^n_+)$, $s > (n+2)/2 + m - 2$, f *is smooth, u satisfies* (5.7), *and*

$$(\partial x_1)^i u\big|_{x_1=0} \in H^{r-i}{}_{ml}(0,0,\tau_0,\xi_0') \text{ for } 0 \le i \le [(m+k)/2] - j - 1.$$

If $u \in H^r{}_{ml}(\Gamma_1\cup\ldots\cup\Gamma_j)$, *then* $u \in H^r{}_{ml}(\Gamma_{j+1}\cup\ldots\cup\Gamma_k)$ *for* $r < 2s - (n+1)/2 - m + 5/2$.

More general boundary conditions may be treated, the global hypotheses may be replaced by local ones, and under stronger assumptions on s similar results hold for the general semilinear, quasilinear or fully nonlinear problem. See Sablé-Tougeron [67] for proofs using the paradifferential calculus.

If the analysis is to be carried out when singularities along glancing null bicharacteristics are present, the corresponding linear regularity theorem must be used. Consider the second order case, with $p_2(t,x,\tau,\xi)$ strictly hyperbolic near the origin. The cotangent space to $\mathbf{R}\times\partial\mathbf{R}^n_+$, given by points $\rho = (t,0,x',\tau,\xi')$, is divided into the elliptic, hyperbolic, and glancing regions E, H, and G, according to whether there are zero, two or one distinct real roots in ξ_1 to the equation $p_2(t,0,x',\tau,\xi_1,\xi') = 0$. If a single real solution exists, then $H_{p_2}x_1(\rho) = 0$, for H_{p_2} the Hamiltonian vector field. If G^k denotes the set of points for which $(H_{p_2})^j x_1(\rho) = 0$, $j < k$, then G is filtered as $G = G^2 \supset G^3 \supset \ldots \supset G^\infty$. The region of second order tangency $G^2\backslash G^3$ is further divided into the diffractive region G_d, for which $(H_{p_2})^2 x_1(\rho) > 0$, and the gliding region G_g, for which $(H_{p_2})^2 x_1(\rho) < 0$.

In order to define the generalized bicharacteristic

$$\gamma(t) = (t, x_1(t), x'(t), \tau(t), \xi'(t))$$

for p_2 through a point $\gamma(t_0)$ of $H\cup G$, it is convenient to use local symplectic coordinates in which

$$p_2(t,x,\tau,\xi) = \xi_1^2 + r(t,x,\tau,\xi')$$

(see Hörmander [37]). If $\gamma(t_0) \in H$, then for t near t_0, $\gamma(t)$ consists of the union of the incoming and reflected null bicharacteristics considered earlier. For $\gamma(t_0) \in G_d$, $\gamma(t)$ is locally the null bicharacteristic for p_2, which is contained in $T^*(\mathbf{R}^{n+1}_+)$ for t in a small deleted neighborhood of t_0. If $\gamma(t_0) \in G_g \cup G^3$, then $d\gamma'/dt(0) = H_{r_0}\gamma'(0)$, for $r_0(t,x',\tau,\xi) = r(t,0,x',\tau,\xi')$ and $\gamma'(t) = (t,x'(t),\tau(t),\xi'(t))$. (See Melrose-Sjöstrand [48].) When $\gamma(t_0) \in G_g$, it follows that $\gamma(t) \in G_g$ for t near t_0, so in particular its (t,x) projection is locally contained in $\mathbf{R} \times \partial\mathbf{R}^n_+$. If $\gamma(t_0) \in G^3 \backslash G^\infty$ it follows that, for t in a small one-sided deleted neighborhood of t_0, either $\gamma(t) \in G_g$ or $\gamma(t)$ is a null bicharacteristic for p_2 in $T^*(\mathbf{R}^{n+1}_+)$. Through points of $G \backslash G^\infty$ the generalized bicharacteristic is locally unique, but an example of Taylor [70] shows that this property does not necessarily hold for the points of infinite order tangency.

In order to establish microlocal regularity for nonlinear problems, David-Williams [30] use the following statement about the linear propagation of singularities, which is a strengthening of the results of Melrose-Sjöstrand [48].

Theorem 5.12. *Let $p(t,x,D)$ be a second order strictly hyperbolic partial differential operator on $\mathbf{R} \times \mathbf{R}^n_+$ such that $\mathbf{R} \times \partial\mathbf{R}^n_+$ is non-characteristic for p_2. Let $f \in H^{r-1,-\infty}{}_{loc}(\mathbf{R} \times \mathbf{R}^n_+)$ for $r \geq 2$, and let $u \in H^{r+1,-\infty}{}_{loc}(\mathbf{R} \times \mathbf{R}^n_+)$ satisfy $p(t,x,D_t,D_x)u = f$, $u|_{x_1=0} = g$. Assume that $\gamma_0 \in T^*(\mathbf{R} \times \mathbf{R}^n_+) \cup \partial T^*(\mathbf{R} \times \mathbf{R}^n_+)$, $f \in H^{r-1}{}_{ml}(\gamma_0)$, $g \in H^r{}_{ml}(\gamma'_0)$ in the case $\gamma_0 \in \partial T^*(\mathbf{R} \times \mathbf{R}^n_+)$, and $u \notin H^r{}_{ml}(\gamma_0)$. Then, either γ_0 is a characteristic point of p_2 contained in $T^*(\mathbf{R} \times \mathbf{R}^n_+)$ or $\gamma_0 \in H \cup G$. If $\gamma_0 \notin G^\infty$, let $\gamma(t)$ be a generalized null bicharacteristic through $\gamma_0 = \gamma(t_0)$. Then there is a neighborhood T of t_0 such that $u \notin H^r{}_{ml}(\gamma(t))$ for any $t \in T$. If $\gamma_0 \in G^\infty$, assume additionally that $u \in H^r{}_{ml}(\{(t,x_1,x',0,1,0)$: $0 < x_1 < \varepsilon\})$. Then there is a generalized null bicharacteristic $\gamma(t)$ through $\gamma_0 = \gamma(t_0)$ and there is a neighborhood T of t_0 such that $u \notin H^r{}_{ml}(\gamma(t))$ for any $t \in T$.*

The proof of Theorem 1.11 may now be imitated to yield the propagation of microlocal singularities in the simple semilinear case.

Theorem 5.13 (David-Williams). *Let $p(t,x,D)$ be a second order strictly hyperbolic partial differential operator on $\mathbf{R} \times \mathbf{R}^n_+$ such that $\mathbf{R} \times \partial\mathbf{R}^n_+$ is non-characteristic for p_2. Suppose that $u \in H^s{}_{loc}(\mathbf{R} \times \mathbf{R}^n_+)$, $s > (n+2)/2$, f is smooth, u satisfies $p(t,x,D)u = f(t,x,u)$, and $u|_{x_1=0} = g$. For $s \leq r < 2s - (n+1)/2 + 1/2$, let $\gamma_0 \in T^*(\mathbf{R} \times \mathbf{R}^n_+) \cup \partial T^*(\mathbf{R} \times \mathbf{R}^n_+)$, let $g \in H^r{}_{ml}(\gamma'_0)$ in the case $\gamma_0 \in \partial T^*(\mathbf{R} \times \mathbf{R}^n_+)$, and suppose that $u \notin H^r{}_{ml}(\gamma_0)$. Then there is a generalized null bicharacteristic $\gamma(t)$ through $\gamma_0 = \gamma(t_0)$ and there is a* *neighborhood T of t_0 such that $u \notin H^r{}_{ml}(\gamma(t))$ for any $t \in T$.*

In the case of second order tangency (corresponding to generalized null bicharacteristics intersecting $\partial T^*(\mathbf{R} \times \mathbf{R}^n_+)$ in the diffractive and gliding sets G_d and G_g described before) this result has been proved for quasilinear and fully nonlinear problems by Leichtnam [43]. Leichtnam also deals with points of higher order tangency, with the order depending on the microlocal regularity index r. His results are extended to the case of any finite order tangency (with generalized null bicharacteristics intersecting $\partial T^*(\mathbf{R} \times \mathbf{R}^n_+)$ in $G \setminus G^\infty$) in Xu [75].

As indicated in Chapter II, for a second order interior problem a stronger statement may be made about the propagation of microlocal regularity: if a nonlinear solution has overall regularity $H^s{}_{loc}(\mathbf{R}^{n+1})$, then additional microlocal regularity is propagated up to order approximately $3s - n$ rather than $2s - n/2$. In Williams [73], [74] it is shown that this result does not in general extend to solutions on regions with boundary; the nonlinear interaction of a pair of microlocal singularities in $\partial T^*(\mathbf{R} \times \mathbf{R}^n_+)$ may result in new singularities which propagate outward. Only under special circumstances on the location of the microlocal singularities does regularity of order better than approximately $2s - n/2$ propagate. The proof of the existence of solutions having singularities of this strength requires the construction of linear examples, and of precise regularity estimates on remainder terms, as in Chapter II. In particular, Williams uses extensions of Theorem 5.12 and Lemma 5.10 to the case of microlocal regularity measured in the Hörmander spaces $H^{r,r'}(\mathbf{R}^n_+)$.

Definition 5.14. For $u \in H^{r,-\infty}(\mathbf{R}^n_+)$ and $(0, x'_0, \xi'_0) \in T^*(\partial \mathbf{R}^n_+)$, we say that $u \in H^{r,r'}{}_{ml}(0, x'_0, \xi'_0)$ if there is a boundary pseudodifferential operator $a(x_1, x', D_{x'})$ of order zero and microlocally elliptic at $(0, x'_0, \xi'_0)$, such that $a(x_1, x', D_{x'})u \in H^{r,r'}{}_{ml}(0, x'_0, \xi'_0)$. For interior points $(x_0, \xi_0) \in T^*(\mathbf{R}^n_+)$, we let $H^{r,r'}{}_{ml}(x_0, \xi_0) = H^{r+r'}{}_{ml}(x_0, \xi_0)$.

The following result on the propagation of singularities as measured in these spaces for solutions to linear strictly hyperbolic problems is proved in Williams [73].

Theorem 5.15. *Let $p(t, x, D)$ be a second order strictly hyperbolic partial differential operator on $\mathbf{R} \times \mathbf{R}^n_+$ such that $\mathbf{R} \times \partial \mathbf{R}^n_+$ is non-characteristic for*

p_2. *Let* $u \in H^{r+1,-\infty}{}_{loc}(\mathbf{R}\times\mathbf{R}^n_+) \cap H^{r+r'+1/2}{}_{ml}(\{(t,x_1,x',0,1,0)\colon 0 < x_1 < \varepsilon\}$, $r \geq 2$, $r' \geq -1$, $f \in H^{r-1,-\infty}{}_{loc}(\mathbf{R}\times\mathbf{R}^n_+)$, *and assume that* $p(t,x,D)u = f$, $u|_{x_1=0} = g$. *Suppose that* $\gamma_0 \in T^*(\mathbf{R}\times\mathbf{R}^n_+) \cup \partial T^*(\mathbf{R}\times\mathbf{R}^n_+)$, $f \in H^{r-1,r'}{}_{ml}(\gamma_0)$, $g \in H^{r+r'-1/2}{}_{ml}(\gamma'_0)$ *in the case* $\gamma_0 \in \partial T^*(\mathbf{R}\times\mathbf{R}^n_+)$, *and* $u \notin H^{r+1,r'-1}{}_{ml}(\gamma_0)$. *Then either* γ_0 *is a characteristic point of* p_2 *in* $T^*(\mathbf{R}\times\mathbf{R}^n_+)$ *or* $\gamma_0 \in H \cup G$. *Let* $0 \leq q \leq r+1$, $q + q' = r + r'$, *and suppose that* $u \notin H^{q,q'}{}_{ml}(\gamma_0)$. *Then there is a generalized null bicharacteristic* $\gamma(t)$ *through* $\gamma_0 = \gamma(t_0)$ *and there is a neighborhood* T *of* t_0 *such that* $u \notin H^{q,q'}{}_{ml}(\gamma(t))$ *for any* $t \in T$.

The spaces $H^{2s-\sigma,-s+\sigma}{}_{loc}(\mathbf{R}^n_+) \cap H^{r,r'}{}_{ml}(x_0,\xi_0)$ are algebras for certain values of the indices. The proof is similar to that of Lemma 5.10.

Lemma 5.16. *Let* $(x_0,\xi'_0) \in T^*(\partial\mathbf{R}^n_+)$. *If* $1/2 < \sigma < s - n/2$ *and* $0 \leq r' \leq \sigma$, *then* $H^{2s-\sigma,-s+\sigma}{}_{loc}(\mathbf{R}^n_+) \cap H^{r,r'}{}_{ml}(x_0,\xi'_0)$ *is an algebra for* $(n+1)/2 < s \leq r < 2s - n/2 - \sigma$.

The usual inductive argument using the algebra and linear regularity results allows the following precise statement about the propagation of microlocal singularites in solutions to simple semilinear equations.

Theorem 5.17 (Williams). *Let* $p(t,x,D)$ *be a second order strictly hyperbolic partial differential operator on* $\mathbf{R}\times\mathbf{R}^n_+$ *such that* $\mathbf{R}\times\partial\mathbf{R}^n_+$ *is non-characteristic for* p_2. *Suppose that* $u \in H^s{}_{loc}(\mathbf{R}\times\mathbf{R}^n_+)$, $s > (n+2)/2$, f *is smooth, and* u *satisfies* $p(t,x,D)u = f(t,x,u)$, $u|_{x_1=0} = g$. *For* $s \leq r$, $0 \leq r' \leq 1/2$, $r + r' < 2s - (n+1)/2 + 1$, *let* $\gamma_0 \in T^*(\mathbf{R}\times\mathbf{R}^n_+) \cup \partial T^*(\mathbf{R}\times\mathbf{R}^n_+)$, *let* $g \in H^r{}_{ml}(\gamma'_0)$ *in the case* $\gamma_0 \in \partial T^*(\mathbf{R}\times\mathbf{R}^n_+)$, *and suppose that* $u \notin H^{r,r'}{}_{ml}(\gamma_0)$. *Then there is a generalized null bicharacteristic* $\gamma(t)$ *through* $\gamma_0 = \gamma(t_0)$ *and there is a neighborhood* T *of* t_0 *such that* $u \notin H^{r,r'}{}_{ml}(\gamma(t))$ *for any* $t \in T$.

As an example, it can be established by propagating regularity in the r' index that the following property holds. Let $p(t,x,D)$ be a second order strictly hyperbolic partial differential operator, with $\mathbf{R}\times\partial\mathbf{R}^n_+$ non-characteristic for p_2, and suppose that all generalized bicharacteristics for p_2 enter $\{t < 0\}$. Let u satisfy

$$p(t,x,D)u = f(t,x,u) \text{ on } \mathbf{R}\times\mathbf{R}^n_+, \quad u(t,0,x') = 0, \tag{5.8}$$

and assume that u has wave front set in $\{t < 0\}$ confined to a single null bicharacteristic which intersects the boundary at $(0,0,0)$. Then the wave front set of u is the generalized null bicharacteristic through $(0,0,0)$. See Chen [27] for the case of transversal intersection. More generally, suppose that $K(t,x)$ is an assignment of cones contained in $\mathbf{R}^{n+1}$ for $(t,x) \in \mathbf{R} \times \mathbf{R}^n_+$ and contained in $\mathbf{R}^n$ for $(t,x) \in \mathbf{R} \times \partial\mathbf{R}^n_+$, satisfying the following properties. See Figure 5.4.

$$\begin{aligned}
&K(t,x) \text{ is proper, } \{K(t,x) + K(t,x)\}^{closure} = K(t,x),\\
&K(t,x_1,x') \cap \{(0,\xi_1,0):\ \xi_1 \in \mathbf{R}\backslash 0\} = \varnothing \ \text{ for } 0 < x_1 < \varepsilon,\\
(5.9)\quad &\text{and if } (t_2,x_2,\tau_2,\xi_2) \text{ comes after } (t_1,x_1,\tau_1,\xi_1) \text{ on a}\\
&\text{generalized bicharacteristic, and } (\tau_1,\xi_1) \in K(t_1,x_1), \text{ then}\\
&(\tau_2,\xi_2) \in K(t_2,x_2).
\end{aligned}$$

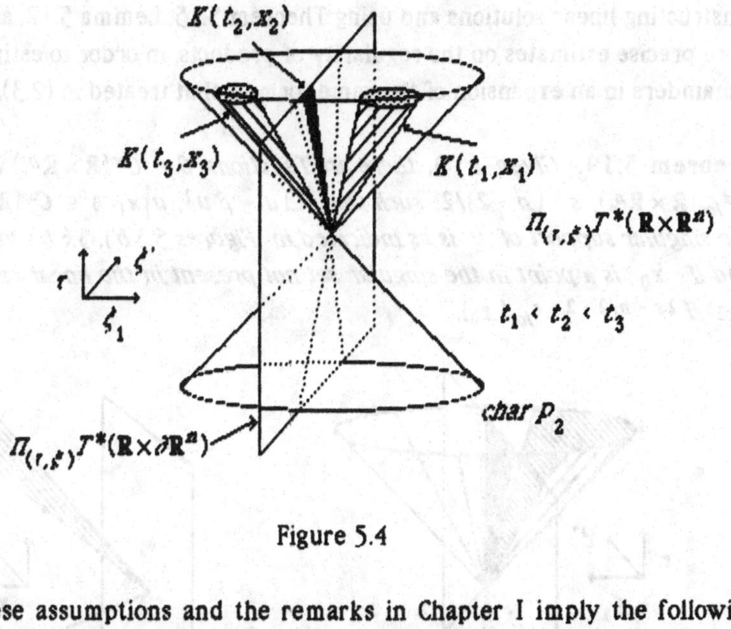

Figure 5.4

These assumptions and the remarks in Chapter I imply the following result. Let $s > (n+2)/2$, $r \geq 2s - 1/2$, $r' \geq 0$, and suppose that

$$\begin{aligned}
u \in\ & H^s{}_{loc}(\mathbf{R} \times \mathbf{R}^n_+) \cap H^{r,s-r}{}_{loc}(\mathbf{R} \times \partial\mathbf{R}^n_+)\\
&\cap H^{r+r'}{}_{ml}\{(t,x,\tau,\xi)\} \text{ for } (t,x) \in \mathbf{R} \times \mathbf{R}^n_+,\ (\tau,\xi) \in K(t,x)^{comp}\\
&\cap H^{r,r'}{}_{ml}\{(t,0,x',\tau,0,\xi')\} \text{ for } (\tau,\xi') \in K(t,0,x')^{comp}.
\end{aligned}$$

If f is analytic, then $f(u)$ has the same properties. From Theorem 5.15 and the relationship between $K(t_2,x_2)$ and $K(t_1,x_1)$ for $t_2 > t_1$, the following result on the absence of nonlinear singularities (Williams [73]) may be deduced.

Theorem 5.18. *Let* $u \in H^s_{loc}(\mathbf{R}\times\mathbf{R}^n_+)$, $s > (n+2)/2$, *satisfy* (5.8), *and let* $K(t,x)$ *be an assignment of cones satisfying* (5.9). *If* $\Pi_{(\tau,\xi)}\,WF(u) \subset K(t,x)$ *for all* $t \le 0$, *then* $\Pi_{(\tau,\xi)}\,WF(u) \subset K(t,x)$ *for all* t.

The cones considered in Chapter II for the interior cases of nonlinear singularities due to crossing and to self-spreading do not have this property. When the semilinear problem (5.8) is considered for a solution with a pair of singularities on incoming null bicharacteristics with projections crossing at the boundary, nonlinear singularities of strength approximately $2s - n/2$ will be created. The following interaction results are proved in Williams [74] by constructing linear solutions and using Theorem 5.15, Lemma 5.12, and other more precise estimates on the regularity of products, in order to estimate the remainders in an expansion of the same form as that treated in (2.3).

Theorem 5.19. *Given* $\varepsilon > 0$, *there are functions* $\beta \in C^\infty(\mathbf{R}\times\mathbf{R}^n_+)$ *and* $u \in H^s_{loc}(\mathbf{R}\times\mathbf{R}^n_+)$, $s > (n+2)/2$, *such that* $\Box u = \beta u^2$, $u|_{x_1=0} \in C^\infty(\mathbf{R}\times\partial\mathbf{R}^n_+)$, *the singular support of* u *is as indicated in Figures* 5.5*b*), 5.6*b*), *or* 5.7*b*), *and if* x_0 *is a point in the singular set not present in the linear case, then* $u \notin H^{2s-n/2+2+\varepsilon}_{loc}(x_0)$.

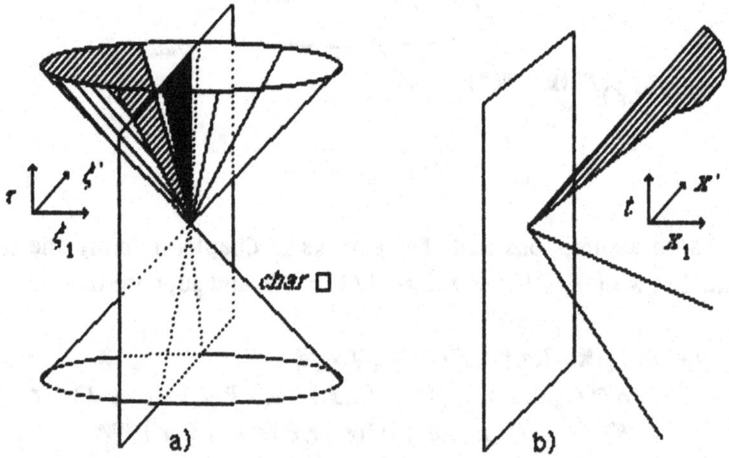

Figure 5.5

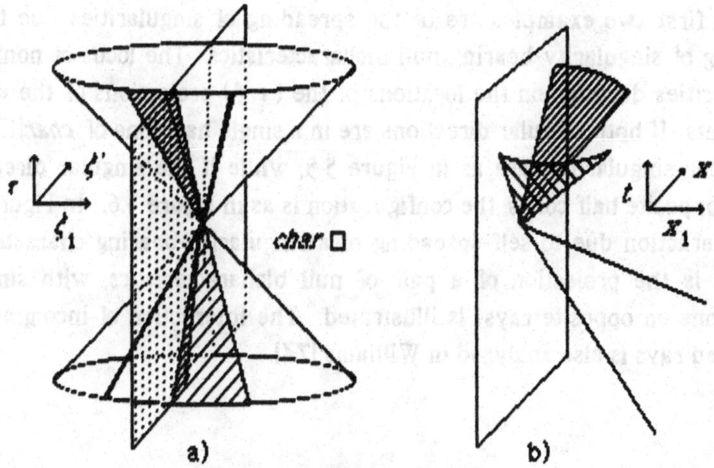

Figure 5.6

The (τ, ξ) projections of the wavefront sets of the functions u and their restrictions to $\mathbf{R} \times \partial \mathbf{R}^n_+$ are shown in Figures 5.5 a), 5.6 a), and 5.7 a). Note that the interaction of a pair of singularities in $\mathbf{R} \times \partial \mathbf{R}^n_+$ is easily seen to yield singularities in new directions which lift to new characteristic directions for $\Box$ in $\mathbf{R} \times \mathbf{R}^n_+$. Thus it is not surprising that arguments like those given in Chapter II lead to the appearance of nonlinear singularities after just a pairwise interaction.

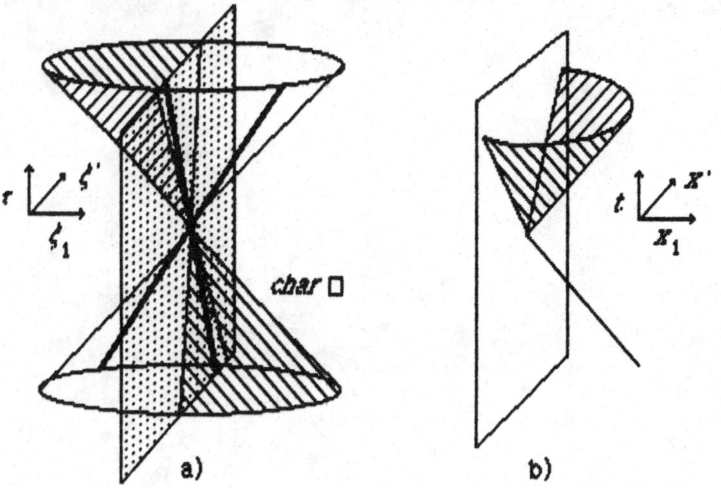

Figure 5.7

The first two examples are of the spreading of singularities due to the crossing of singularity-bearing null bicharacteristics. The locus of nonlinear singularities depends on the locations of the (x, ξ) projections of the wave-front sets. If both singular directions are in a single half cone of $char(\Box)$, the nonlinear singularities are as in Figure 5.5, while if the singular directions are in opposite half cones, the configuration is as in Figure 5.6. In Figure 5.7, the interaction due to self-spreading of a singularity-bearing characteristic (which is the projection of a pair of null bicharacteristics, with singular directions on opposite rays) is illustrated. The interaction of incoming and reflected rays is also analyzed in Williams [74].

Chapter VI. Conormal Waves on Domains with Boundary

It is natural to consider the propagation of more restricted types of regularity for solutions to strictly hyperbolic problems on domains with boundary, just as in the case of the interior problem considered in Chapters III and IV. From Theorems 5.11 and 5.13, if $u \in H^s{}_{loc}(\mathbf{R}\times\mathbf{R}^n{}_+)$, with $s > (n+2)/2 + m - 2$, is a solution to (5.2), microlocal regularity of type H^r for u propagates along reflected families of non-grazing null bicharacteristics for $r < 2s - (n+2)/2 - m + 5/2$. The same result holds along (possibly grazing) generalized null bicharacteristics in the case of a second order operator. If one is interested in solutions with singularities for which higher order microlocal regularity is propagated (up to H^∞), it is again appropriate to study functions conormal with respect to families of characteristic hypersurfaces.

The simplest such boundary problem is one for a second order semilinear equation, with a solution having conormal singularities in the past across a single smooth characteristic hypersurface Σ^-. In the linear case, the wavefront set of such a solution would in the past be contained in the conormal bundle $N^*(\Sigma^-)$; the linear analysis is simplest when none of the null bicharacteristics whose union forms $N^*(\Sigma^-)$ is in the glancing set G defined before Theorem 5.8. Therefore, we assume that the null bicharacteristics intersect the boundary transversally; in particular, the intersection of Σ^- with the boundary is then a smooth submanifold Δ of codimension one in the boundary, and the union of the reflected null bicharacteristics through $N^*(\Delta)$ is the conormal bundle $N^*(\Sigma^+)$ to a (locally) smooth characteristic hypersurface Σ^+ which intersects Σ^- transversally at Δ. After a smooth local change of coordinates, it may be assumed that the region in question is $\mathbf{R}\times\mathbf{R}^n{}_+$ and the hypersurfaces are given by $\{t = \varphi_-(x_1,x')\}$ and $\{t = \varphi_+(x_1,x')\}$, with smooth functions $\varphi_\pm$ satisfying $\varphi_-(0,x') = \varphi_+(0,x')$, $(\partial\varphi_-/\partial x_1)(0,x') \neq (\partial\varphi_+/\partial x_1)(0,x')$. If we set

$$s = 2t - (\varphi_+(x_1,x') + \varphi_-(x_1,x')),\ y_1 = \varphi_+(x_1,x') - \varphi_-(x_1,x'),\ y' = x',$$

then in the new coordinates the boundary is $\{y_1 = 0\}$ and the characteristic hypersurfaces correspond to $\{s = -y_1\}$ and $\{s = y_1\}$. After changing notation back to the variables (t,x), we make the following assumptions; see Figure 6.1. The global hypotheses may be replaced by local ones through the use of finite propagation speed.

(6.1) $\Sigma^- = \{t = -x_1\}$ and $\Sigma^+ = \{t = x_1\}$ are smooth characteristic hypersurfaces for the second order strictly hyperbolic operator $p_2(t,x,D)$ on $\mathbf{R}\times\mathbf{R}^n_+$, and $\mathbf{R}\times\partial\mathbf{R}^n_+$ is non-characteristic for p_2.

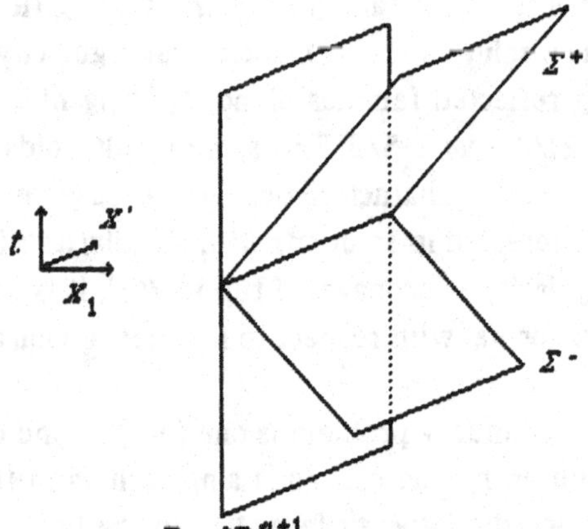

Figure 6.1

In order to use a commutator argument in a semilinear problem, we consider

$$\mathcal{M} = \{\text{smooth vector fields on } \mathbf{R}\times\mathbf{R}^n_+ \text{ simultaneously tangent to } \Sigma^-, \Sigma^+, \text{ and } \mathbf{R}\times\partial\mathbf{R}^n_+\}.$$

The conormal space is defined in the natural fashion.

Definition 6.1. For a pair of smooth hypersurfaces $\Sigma^-,\Sigma^+ \subset \mathbf{R}\times\mathbf{R}^n_+$ which intersect transversally at $\mathbf{R}\times\partial\mathbf{R}^n_+$, $u \in H^s_{loc}(\mathbf{R}\times\mathbf{R}^n_+)$ is said to be conormal with respect to the family $\{\Sigma^-, \Sigma^+, \mathbf{R}\times\partial\mathbf{R}^n_+\}$ if $M_1 \cdots M_j u \in H^s_{loc}(\mathbf{R}\times\mathbf{R}^n_+)$ for $M_1, \ldots, M_j \in \mathcal{M}$ as given above. If this property holds for all $j \leq k$, u is said to be conormal of degree k with respect to $\{\Sigma^-, \Sigma^+, \mathbf{R}\times\partial\mathbf{R}^n_+\}$, written $u \in N^{s,k}(\Sigma^-, \Sigma^+, \mathbf{R}\times\partial\mathbf{R}^n_+)$.

Exactly as in Lemma 3.2, for $s > (n+1)/2$ these spaces are algebras invariant under the action of smooth functions. A convenient set of generators of $\mathcal{M}$ is determined in coordinates chosen so that the surfaces are in the standard form (6.1).

Lemma 6.2. *$\mathcal{M}$ is generated over $C^\infty(\mathbf{R}\times\mathbf{R}^n_+)$ by the vector fields*

$$M_0 = t\partial_t + x_1\partial_{x_1},\ M_1 = x_1(t+x_1)(\partial_t + \partial_{x_1}),\ M_j = \partial_{x_j},\ j = 2, \ldots, n.$$

M_1 may be equivalently replaced with $M_1' = x_1(t - x_1)(\partial_t - \partial x_1)$.

Proof. Each of these vector fields is easily seen to be an element of $\mathcal{M}$. Moreover, every smooth vector field on $\mathbf{R}\times\mathbf{R}^n_+$ can be written as

$$M = \alpha(\partial_t + \partial x_1) + \beta(\partial_t - \partial x_1) + \gamma_2 M_2 + \ldots + \gamma_n M_n.$$

If M is tangent to Σ^- and Σ^+, then there are smooth functions α' and β' with $\alpha = (t+x_1)\alpha'$ and $\beta = (t-x_1)\beta'$. If, in addition, M is tangent to $\{x_1 = 0\}$, then x_1 divides $\alpha' - \beta'$, so there is a smooth function γ_1 such that

$$\begin{aligned}\alpha(\partial_t + \partial x_1) + \beta(\partial_t - \partial x_1) &= \beta'\{(t+x_1)(\partial_t + \partial x_1) + (t-x_1)(\partial_t - \partial x_1)\} + \gamma_1 M_1\\ &= 2\beta' M_0 + \gamma_1 M_1.\end{aligned}$$

Finally, M_1' may be used instead, since $M_1' = 2x_1 M_0 - M_1$. Q.E.D.

As is easily checked, the generators given above satisfy the following commutation relations.

$$\begin{aligned}&[M_0, M_1] = M_1,\ [M_0, M_1'] = M_1',\\ (6.2)\qquad &[M_1, M_1'] = (t+x_1)M_1' + (t-x_1)M_1,\\ &\text{and all other commutators are zero.}\end{aligned}$$

In order to commute these vector fields past the operator p_2, it is necessary to find a standard form for the symbol.

Lemma 6.3. *If p_2 satisfies (6.1), then locally near the origin, after division by a nonzero factor, there are polynomials $r(t,x,\tau,\xi)$ and $q_j(t,x,\tau,\xi)$ of order one such that $p_2(t,x,\tau,\xi) = (\tau^2 - \xi_1^2) + (t\tau + x_1\xi_1)r + \xi_2 q_2 + \dots + \xi_n q_n$.*

Proof. We write

$$p_2 = a(\tau + \xi_1)^2 + b(\tau - \xi_1)^2 + c(\tau^2 - \xi_1^2) + \xi_2 q_2 + \dots + \xi_n q_n.$$

Since $\{t = -x_1\}$ and $\{t = x_1\}$ are characteristic for p_2, it follows that there are smooth functions a' and b' with $a = (t + x_1)a'$ and $b = (t - x_1)b'$. Furthermore, $(t + x_1)(\tau + \xi_1) = -(t - x_1)(\tau - \xi_1) + 2(t\tau + x_1\xi_1)$, and thus

$$\begin{aligned} a(\tau + \xi_1)^2 + b(\tau - \xi_1)^2 = &-a'(t - x_1)(\tau - \xi_1)(\tau + \xi_1) \\ &- b'(t + x_1)(\tau + \xi_1)(\tau - \xi_1) + (t\tau + x_1\xi_1)r. \end{aligned}$$

Therefore, $p_2 = c'(\tau^2 - \xi_1^2) + (t\tau + x_1\xi_1)r + \xi_2 q_2 + \dots + \xi_n q_n$. The strict hyperbolicity of p_2 implies that $c' \neq 0$ near the origin. Q.E.D.

The desired form for the commutators of the generating vector fields with p_2 may be deduced from these expressions.

Lemma 6.4. *There are smooth functions α_j and first order operators $R_{j,k}$ and R_j such that the commutators satisfy $[p_2(t,x,D), M_j] = \alpha_j p_2(t,x,D) + R_{j,0}M_0 + R_{j,1}M_1 + \dots + R_{j,n}M_n + R_j$, for $j = 0, 1, \dots, n$.*

Proof. From Lemma 6.3 and (6.2), it follows that there are first order operators $R_{j,k}$ and R_j such that

$$[p_2(t,x,D), M_j] = [\partial_t^2 - \partial_{x_1}^2, M_j] + R_{j,0}M_0 + R_{j,1}M_1 + \dots + R_{j,n}M_n + R_j .$$

Moreover, $[p_2(t,x,D), M_0] = 2(\partial_t^2 - \partial_{x_1}^2)$, $[p_2(t,x,D), M_j] = 0$ for $j = 2, \dots, n$, and

$$[p_2(t,x,D), M_1] = 2(x + t)(\partial_t^2 - \partial_{x_1}^2) - 2(\partial_t + \partial_{x_1})M_0 + R_1$$

for a first order operator R_1. Since $\partial_t^2 - \partial x_1^2 = p_2(t,x,D)$ modulo a term of the form $R_{j,0}M_0 + R_{j,1}M_1 + \ldots + R_{j,n}M_n + R_j$, the desired expression is obtained. Q.E.D.

These results may be used to establish the propagation of conormal regularity with respect to hypersurfaces as in (6.1) for solutions to simple semilinear equations. In particular, the singular support of a solution u will be as indicated in Figure 6.1. The global hypotheses in the following statement may be replaced by appropriate local ones, by means of finite propagation speed and the interior conormal regularity theorem already proved.

Theorem 6.5 (Beals-Métivier). *Let $p(t,x,D)$ be a second order strictly hyperbolic operator on $\mathbf{R}\times\mathbf{R}^n_+$ such that $\mathbf{R}\times\partial\mathbf{R}^n_+$ is non-characteristic for p_2. Let Σ^- and Σ^+ be a reflected pair of smooth characteristic hypersurfaces which intersect transversally at $\mathbf{R}\times\partial\mathbf{R}^n_+$ in $\{t \geq 0\}$. Suppose that $u \in H^s_{loc}(\mathbf{R}\times\partial\mathbf{R}^n_+)$, $s > (n+1)/2$, f is a smooth function, and u satisfies $p(t,x,D)u = f(t,x,u)$, $u(t,0,x') = 0$. If $u \in N^{s,k}(\Sigma^- \cap \{t < -\varepsilon\})$ for some $\varepsilon > 0$, then $u \in N^{s,k}(\Sigma^-, \Sigma^+, \mathbf{R}\times\partial\mathbf{R}^n_+)$.*

Proof. Suppose inductively that $u \in N^{s,j-1}(\Sigma^-, \Sigma^+, \mathbf{R}\times\partial\mathbf{R}^n_+)$ for some j with $1 \leq j \leq k$. Then $f(t,x,u) \in N^{s,j-1}(\Sigma^-, \Sigma^+, \mathbf{R}\times\partial\mathbf{R}^n_+)$ by the algebra property. Set $U = (M_0^{\alpha_0} \cdots M_n^{\alpha_n} u\colon |\alpha| \leq j)$. Then $U(t,0,x') = 0$, because all of the vector fields $M_0, \ldots, M_n$ are tangent to $\{x_1 = 0\}$. (This is the reason for including $\mathbf{R}\times\partial\mathbf{R}^n_+$ in the family of hypersurfaces used to define conormal regularity.) Moreover, it is an easy consequence of Lemma 6.4 that there is a system of first order operators $r_1(t,x,D)$ and a vector of operators $q_k(t,x,M)$ of order k acting on $M_0, \ldots, M_n$ such that

$$p(t,x,D)U = r_1(t,x,D)U + q_k(t,x,M)f(t,x,u).$$

By the inductive hypothesis,

$$(p - r_1)(t,x,D)U \in H^{s-1}_{loc}(\mathbf{R}\times\mathbf{R}^n_+),$$
$$U \in H^s_{loc}(\{t < -\varepsilon\}\times\mathbf{R}^n_+), \text{ and } U(t,0,x') = 0,$$

so the linear energy inequality for problems on domains with boundary (see Chazarain-Piriou [21]) applied to this system with diagonal principal part implies that $U \in H^s_{loc}(\mathbf{R}\times\mathbf{R}^n_+)$. Therefore $u \in N^{s,j}(\Sigma^-, \Sigma^+, \mathbf{R}\times\partial\mathbf{R}^n_+)$, and the result follows. Q.E.D.

More general boundary conditions and higher order operators are treated in Beals-Métivier [11], [12], again under the assumption that only transversally intersecting characteristic hypersurfaces are present in the reflected family. The precise hypotheses are as follows.

(6.3) $p(t,x,D)$ is a strictly hyperbolic operator of order m on $\mathbf{R}\times\mathbf{R}^n_+$, $\mathbf{R}\times\partial\mathbf{R}^n_+$ is non-characteristic for p_m, and $B = (B_j)_{j=0,\ldots,\mu}$ is a collection of operators on $\mathbf{R}\times\partial\mathbf{R}^n_+$ satisfying the uniform Lopatinski condition with respect to p_m.

See Chazarain-Piriou [21] for the definition of the uniform Lopatinski condition. It insures that the solution to a linear problem of the form

$$p(t,x,D)u = f(t,x) \text{ on } \mathbf{R}\times\mathbf{R}^n_+,\ Bu\big|_{x_1=0} = g(t,x'),$$

satisfies the appropriate energy estimate.

If Σ_1 is a smooth characteristic hypersurface for p_m, intersecting $\Sigma_0 = \mathbf{R}\times\partial\mathbf{R}^n_+$ transversally in the smooth submanifold Δ, for $(t_0,0,x'_0)\in\Delta$ let $(\tau_0,\xi_{1,1},\xi'_0)$ denote a unit conormal to Σ_1. The nongrazing hypothesis may be stated as follows:

(6.4) the real roots of $p_m(t_0,0,x'_0,\tau_0,\xi_1,\xi'_0) = 0$ are simple for $(t_0,0,x'_0)\in\Delta$.

Under this assumption, if there are k real roots, then there are unit vectors $(\tau_0,\xi_{1,j},\xi'_0)$, $1\le j\le k$, which can be chosen to vary smoothly with (t_0,x'_0). For example, the vectors $(\tau_0,\xi_{1,j},\xi'_0)$ corresponding to fourth order operators for which $k=2$ and $k=4$ are illustrated in Figure 5.2. The null bicharacteristics for p_m passing through $(t_0,0,x'_0,\tau_0,\xi_{1,j},\xi'_0)$ have (t,x) projections which constitute a family of smooth characteristic hypersurfaces $(\Sigma_1,\ldots,\Sigma_k)$; the pairwise intersections are transverse.

If only two hypersurfaces (Σ_1,Σ_2) are present in the reflected family, the proof of Theorem 6.5 may be adapted to the case of a higher order operator. If coordinates are chosen so that the surfaces are as in (6.1), the same vector fields M_j as in Lemma 6.2 may be used to define conormal regularity.

However, a good expression for the commutators $[p_m(t,x;D), M_j]$ is only obtained microlocally on sets which avoid the $m-2$ roots in of the equation $p_m(0,0,x',\tau,\xi_1,\xi') = 0$ distinct from $(0,0,x',\tau,\tau,0)$ and $(0,0,x',\tau,-\tau,0)$. Indeed, microlocally near those two roots, p_m may be written as the product of a term like p_2 in Lemma 6.3 and an elliptic term e_{m-2} of order $m-2$. The calculus of pseudodifferential operators and division by the elliptic factor then imply that the analogue of Lemma 6.4 holds on this microlocal region. Microlocally near the other $m-2$ roots this expression will not in general hold. In addition, since on a region with boundary we are for the most part restricted to using boundary pseudodifferential operators as cutoffs in the microlocal calculus, the simple commutator argument works only on a microlocal neighborhood with (τ,ξ) projection of the form $\{(\tau,\xi_1\xi'): |\xi'| < \varepsilon|\tau|\}$. With tangential microlocalization essential, it is again necessary to use the spaces $H^{s,s'}{}_{loc}(\mathbf{R}\times\mathbf{R}^n{}_+)$, since the remainder terms in the tangential microlocal calculus only improve regularity in the second index. Near $\{(\tau,\xi_1,\xi'): |\xi'| < \varepsilon|\tau|\}$, the commutator argument as in the proof of Theorem 6.5 may be applied; for $\{(\tau,\xi_1,\xi'): |\xi'| \geq \varepsilon|\tau|\}$, an inductive argument using the partial hypoellipticity of p_m with respect to x_1 yields the desired estimates. For details, see Beals-Métivier [11].

When more than two characteristic hypersurfaces $\{\Sigma_1,\ldots,\Sigma_k\}$ are contained in the reflected family determined by (6.4), as in Chapter III it is no longer possible to define a useful conormal space by considering all vector fields simultaneously tangent to $\Sigma_0,\Sigma_1,\ldots,\Sigma_k$; these vector fields would vanish of too high a degree at Δ for commutators with p_m to have suitable expressions. The collection of hypersurfaces satisfies the hypotheses for the conormal space of Definition 3.7; since that definition involves microlocalization we begin by extending the hypersurfaces and functions under consideration to all of $\mathbf{R}\times\mathbf{R}^n$. On the other hand, if u satisfies

$$p(t,x,D)u = f(t,x) \text{ on } \mathbf{R}\times\mathbf{R}^n{}_+,\ Bu\big|_{x_1=0} = g(t,x'),$$

it is not necessarily true that it is possible to extend u across the boundary in a fashion which preserves the regularity under consideration. Thus it is important to have another characterization of the spaces in Definition 3.7.

Lemma 6.6. *Let* $\{\Sigma_0 = \mathbf{R}\times\partial\mathbf{R}^n{}_+,\Sigma_1,\ldots,\Sigma_k\}\subset\mathbf{R}^{n+1}$ *be a family of smooth hypersurfaces intersecting pairwise transversally in the single codimension two submanifold* Δ. *Then* $u\in N^{s,j}(\Sigma_0,\Sigma_1,\ldots,\Sigma_k)$ *if and only if there are* *functions* $u_0 \in N^{s,j}(\Sigma_0,\Delta)$, $u_j \in N^{s,j}(\Sigma_0,\Sigma_{k'})$ *for* $1 \le k' \le k$, *such that* $u = u_0 + u_1 + \ldots + u_k$.

Here the spaces $N^{s,j}(\Sigma_0,\Sigma_{k'})$ are determined as in Definition 3.4, while $u \in N^{s,j}(\Sigma_0,\Delta)$ means that $M_1 \cdots M_{j'} u \in H^s_{loc}(\mathbf{R}^{n+1})$ for all smooth vector fields $M_1, \ldots, M_{j'}$ simultaneously tangent to Σ_0 and Δ, for $j' \le j$.

Proof. Let $\chi_{k'}(t,x,D)$ be a family of smooth pseudodifferential operators of order zero with disjoint conic supports, $\chi_{k'} = 1$ on a conic neighborhood of $N^*(\Sigma_{k'})$, with $\chi_0 + \chi_1 + \ldots + \chi_k = 1$. If M is simultaneously tangent to Σ_0 and $\Sigma_{k'}$ (or Σ_0 and Δ if $k' = 0$) then $M\chi_{k'}(t,x,D)$ has principal symbol which vanishes on $N^*(\Sigma_0) \cup N^*(\Sigma_1) \cup \ldots \cup N^*(\Sigma_k) \cup N^*(\Delta)$. It follows for $u \in N^{s,j}(\Sigma_0,\Sigma_1,\ldots,\Sigma_k)$ that

$$\chi_{k'}(t,x,D)u \in N^{s,j}(\Sigma_0,\Sigma_{k'}),\ 1 \le k' \le k, \text{ and}$$
$$\chi_0(t,x,D)u \in N^{s,j}(\Sigma_0,\Delta).$$

On the other hand, the argument of Bony [17] indicated after Definition 3.7 establishes that $N^{s,j}(\Sigma_0,\Delta) \subset N^{s,j}(\Sigma_0,\Sigma_1,\ldots,\Sigma_k)$ and that $N^{s,j}(\Sigma_0,\Sigma_{k'}) \subset N^{s,j}(\Sigma_0,\Sigma_1,\ldots,\Sigma_k)$ for $1 \le k' \le k$. Q.E.D.

The advantage to the characterization given by Lemma 6.6 is that the definition of conormal can be made purely in terms of vector fields, and hence makes sense on the half space $\mathbf{R} \times \mathbf{R}^n_+$.

Definition 6.7. Let $N^{s,j}_+(\Sigma_0,\Sigma_{k'})$ be the set of all $u \in H^s_{loc}(\mathbf{R} \times \mathbf{R}^n_+)$ such that $M_1 \cdots M_{j'} u \in H^s_{loc}(\mathbf{R} \times \mathbf{R}^n_+)$ for all smooth vector fields $M_1, \ldots, M_{j'}$ which are simultaneously tangent to Σ_0 and $\Sigma_{k'}$, for $j' \le j$. Similarly, $u \in N^{s,j}_+(\Sigma_0,\Delta)$ means that $M_1 \cdots M_{j'} u \in H^s_{loc}(\mathbf{R} \times \mathbf{R}^n_+)$ for all smooth vector fields $M_1, \ldots, M_{j'}$ which are simultaneously tangent to Σ_0 and Δ.

Elements of $N^{s,j}_+(\Sigma_0,\Sigma_{k'})$ and $N^{s,j}_+(\Sigma_0,\Delta)$ are extendible to elements of $N^{s,j}(\Sigma_0,\Sigma_{k'})$ and $N^{s,j}(\Sigma_0,\Delta)$.

Lemma 6.8. *Let* $u \in N^{s,j}_+(\Sigma_0,\Sigma_{k'})$. *There exists* $v \in N^{s,j}(\Sigma_0,\Sigma_{k'})$, *with norm locally comparable to the norm of* u, *such that* $v|_{\mathbf{R}\times\mathbf{R}^n_+} = u$ *and* $M_1 \cdots M_{j'} v|_{\mathbf{R}\times\mathbf{R}^n_+} = M_1 \cdots M_{j'} u$ *for all smooth vector fields* $M_1, \ldots, M_{j'}$ *which are simultaneously tangent to* Σ_0 *and* $\Sigma_{k'}$, $j' \le j$. *The analogous state-*

ment holds for $u \in N^{s,l}{}_+(\Sigma_0, \Delta)$, *with* $v \in N^{s,l}(\Sigma_0, \Delta)$.

Proof. We choose coordinates (x_0, x_1, x') in which Σ_0 is given by $\{x_1 = 0\}$ and $\Sigma_{k'}$ is given by $\{x_0 = 0\}$. A set of generators of the vector fields simultaneously tangent to Σ_0 and $\Sigma_{k'}$ is $\mathcal{M} = \{x_0 \partial x_0, x_1 \partial x_1, \partial x_2, \ldots, \partial x_n\}$. Fix $l \geq s + j$, and let constants α_k be chosen with

$$\sum_{k=0}^{l} \alpha_k \left(\frac{-1}{k+1}\right)^m = 1 \text{ for } m = 0, \ldots, l.$$

Set

$$\begin{aligned} Eu(x) &= u(x), & x_1 \geq 0, \\ &= \sum_{k=0}^{l} \alpha_k \, u(x_0, \frac{-x_1}{k+1}, x'), & x_1 < 0. \end{aligned}$$

Clearly, E is a continuous extension from $H^s{}_{loc}(\mathbf{R} \times \mathbf{R}^n{}_+)$ into $H^s{}_{loc}(\mathbf{R} \times \mathbf{R}^n)$. Furthermore, $MEu = EMu$ for $M \in \mathcal{M}$, and hence $Eu \in N^{s,l}(\Sigma_0, \Sigma_{k'})$ for $u \in N^{s,l}{}_+(\Sigma_0, \Sigma_{k'})$.

On the other hand, suppose that $u \in N^{s,l}{}_+(\Sigma_0, \Delta)$. If coordinates are chosen in which Σ_0 is given by $\{x_1 = 0\}$ and Δ is given by $\{x_0 = 0, x_1 = 0\}$, then a set of generators of the vector fields simultaneously tangent to Σ_0 and Δ is $\mathcal{M}_0 = \mathcal{M} \cup \{x_1 \partial x_0\}$. For $0 \leq m \leq l$, set

$$\begin{aligned} E_m u(x) &= u(x), & x_1 \geq 0, \\ &= \sum_{k=0} \alpha_k \left(\frac{-1}{k+1}\right)^m u(x_0, \frac{-x_1}{k+1}, x'), & x_1 < 0. \end{aligned}$$

Each E_m is a continuous extension from $H^s{}_{loc}(\mathbf{R} \times \mathbf{R}^n{}_+)$ into $H^s{}_{loc}(\mathbf{R} \times \mathbf{R}^n)$, and for $M \in \mathcal{M}$, $ME_m u = E_m Mu$. Since $(x_1 \partial x_0) E_m u = E_{m-1}(x_1 \partial x_0) u$, it follows that $E_j u \in N^{s,l}(\Sigma_0, \Sigma_{k'})$ for $u \in N^{s,l}{}_+(\Sigma_0, \Sigma_{k'})$. Q.E.D.

As a consequence of Lemmas 6.6 and 6.8, it is natural make the definition of $N^{s,l}{}_+(\Sigma_0, \Sigma_1, \ldots, \Sigma_k)$ in terms of restrictions.

Definition 6.9. Let $N^{s,l}{}_+(\Sigma_0, \Sigma_1, \ldots, \Sigma_k)$ denote the set of all restrictions to $\mathbf{R} \times \mathbf{R}^n{}_+$ of functions $u = u_0 + u_1 + \ldots + u_k$ such that $u_0 \in N^{s,l}(\Sigma_0, D)$ and $u_i \in N^{s,l}(\Sigma_0, \Sigma_i)$ for $1 \leq i \leq k$.

It follows immediately from Definitions 6.7 and 6.9 and Lemma 6.6 that

(6.5) $$N^{s,j}_+(\Sigma_0,\Sigma_1,\dots,\Sigma_k) = N^{s,j}_+(\Sigma_0,\Delta) + N^{s,j}_+(\Sigma_0,\Sigma_1) + \dots + N^{s,j}_+(\Sigma_0,\Sigma_k).$$

Moreover, the algebra property for such functions on the half-space follows from the corresponding property on $\mathbf{R}\times\mathbf{R}^n$.

Lemma 6.10. *If* $u \in N^{s,j}_+(\Sigma_0,\Sigma_1,\dots,\Sigma_k)$ *for* $s > (n+1)/2$ *and* f *is smooth, then* $f(t,x,u) \in N^{s,j}_+(\Sigma_0,\Sigma_1,\dots,\Sigma_k)$.

The traces of such conormal functions on the boundary $\{x_1 = 0\}$ are conormal with respect to the smooth submanifold Δ, as in Definition 3.1. When the strictly hyperbolic equation under consideration is reduced to a first order system, the components corresponding to elliptic equations will in general have traces with lower Sobolev regularity, unlike the components corresponding to hyperbolic equations as in Lemma 5.7.

Lemma 6.11. *If* $u(t,x_1,x') \in N^{s,j}_+(\Sigma_0,\Sigma_1,\dots,\Sigma_k)$ *for* $s > 1/2$, *then* $u(t,0,x') \in N^{s-1/2,j}(\Delta)$.

Proof. The vector fields used to define $N^{s,j}_+(\Sigma_0,\Sigma_k)$ and $N^{s,j}_+(\Sigma_0,\Delta)$ are all tangent to $\Sigma_0 = \{x_1 = 0\}$, and their restrictions to $\{x_1 = 0\}$ are all tangent to Δ. The result therefore follows from Definition 6.9 and induction on j, since the case $j = 0$ is the usual restriction theorem for Sobolev spaces. (See, for example, Treves [71].) Q.E.D.

For tangential microlocalizations, the coordinates (t,x_1,x') will be fixed in order to define the Hörmander spaces $H^{s,s'}_{loc}(\mathbf{R}\times\mathbf{R}^n_+)$. In addition, it is necessary to consider regularity with respect to the one further vector field which is tangent to the boundary.

Definition 6.12. $\mathcal{D}^{s,j} = \{u\colon (x_1\partial_{x_1})^{j'}u \in H^{s,\infty}_{loc}(\mathbf{R}\times\mathbf{R}^n),\ 0 \le j' \le j\}$, $\mathcal{D}'^{s,j} = \{u\colon (x_1\partial_{x_1})^{j'}u \in H^{s,-\infty}_{loc}(\mathbf{R}\times\mathbf{R}^n),\ 0 \le j' \le j\}$, and on the half-space $\mathbf{R}\times\mathbf{R}^n_+$, $\mathcal{D}^{s,j}_+$ and $\mathcal{D}'^{s,j}_+$ are defined similarly.

The following inclusion relations are an immediate consequence:

$$\mathcal{D}^{s,j} \subset N^{s,j}(\Sigma_0,\Delta) \subset N^{s,j}(\Sigma_0,\Sigma_1,\ldots,\Sigma_k) \subset \mathcal{D}'^{s,j},$$
(6.6)
$$\mathcal{D}^{s,j}_+ \subset N^{s,j}_+(\Sigma_0,\Delta) \subset N^{s,j}_+(\Sigma_0,\Sigma_1,\ldots,\Sigma_k) \subset \mathcal{D}'^{s,j}_+ .$$

Moreover, the infinitely smoothing remainder terms in the tangential pseudodifferential calculus are easily seen to map $\mathcal{D}'^{s,j}$ into $\mathcal{D}^{s,j}$, and similarly $\mathcal{D}'^{s,j}_+$ into $\mathcal{D}^{s,j}_+$.

Lemma 6.13. *Let $p(t,x,D_t,D_{x'})$ be a boundary pseudodifferential operator. If p is of order zero, then $p(t,x,D_t,D_{x'})$ maps the spaces $N^{s,j}(\Sigma_0,\Sigma_{k'})$, $N^{s,j}(\Sigma_0,\Delta)$, $N^{s,j}(\Sigma_0,\Sigma_1,\ldots,\Sigma_k)$, $\mathcal{D}^{s,j}$, and $\mathcal{D}'^{s,j}$ into themselves. If p is of order $-\infty$, then $p(t,x,D_t,D_{x'})$ maps $\mathcal{D}'^{s,j}$ into $\mathcal{D}^{s,j}$. Analogous results hold for $N^{s,j}_+(\Sigma_0,\Sigma_{k'})$, $N^{s,j}_+(\Sigma_0,\Delta)$, $N^{s,j}_+(\Sigma_0,\Sigma_1,\ldots,\Sigma_k)$, $\mathcal{D}^{s,j}_+$, and $\mathcal{D}'^{s,j}_+$.*

Proof. Since

$$[x_1\partial x_1, p(t,x,D_t,D_{x'})] = q(t,x,D_t,D_{x'}), \text{ with}$$
$$q(t,x,\tau,\xi') = x_1\,\partial p/\partial x_1(t,x,\tau,\xi'),$$

it follows that $[M, p(t,x,D_t,D_{x'})]$ is a boundary pseudodifferential operator of the same order as p for any vector field M which is tangent to Σ_0. If p has order zero, it follows by induction on j that the spaces $N^{s,j}(\Sigma_0,\Sigma_{k'})$, $N^{s,j}(\Sigma_0,\Delta)$, $N^{s,j}(\Sigma_0,\Sigma_1,\ldots,\Sigma_k)$, $\mathcal{D}^{s,j}$, and $\mathcal{D}'^{s,j}$ are preserved under the action of $p(t,x,D_t,D_{x'})$. If p has order $-\infty$, it maps $H^{s,-\infty}{}_{loc}(\mathbf{R}\times\mathbf{R}^n)$ into $H^{s,\infty}{}_{loc}(\mathbf{R}\times\mathbf{R}^n)$, and then $\mathcal{D}'^{s,j}$ into $\mathcal{D}^{s,j}$ by induction on j. The proofs on $\mathbf{R}\times\mathbf{R}^n_+$ follow from Lemma 6.8 and Definition 6.9 in the same fashion. Q.E.D.

It is a consequence of (6.6) and Lemma 6.13 that functions in the spaces $N^{s,j}(\Sigma_0,\Sigma_1,\ldots,\Sigma_k)$ and $N^{s,j}_+(\Sigma_0,\Sigma_1,\ldots,\Sigma_k)$ can be tangentially microlocalized in a well defined fashion.

Definition 6.14. For $u \in \mathcal{D}'^{s,j}$ and $(t,x,\tau,\xi') \in \mathbf{R}^{n+1}\times\mathbf{R}^n\backslash 0$, we say that $u \in N^{s,j}(\Sigma_0,\Sigma_1,\ldots,\Sigma_k)$ tangentially microlocally at (t,x,τ,ξ') if there is a boundary pseudodifferential operator $p(t,x,D_t,D_{x'})$ microlocally elliptic at (t,x,τ,ξ') such that $p(t,x,D_t,D_{x'})u \in N^{s,j}(\Sigma_0,\Sigma_1,\ldots,\Sigma_k)$. Analogous definitions are made for tangential microlocalization in the spaces $N^{s,j}(\Sigma_0,\Sigma_{k'})$, $N^{s,j}(\Sigma_0,\Delta)$, $N^{s,j}_+(\Sigma_0,\Sigma_1,\ldots,\Sigma_k)$, $N^{s,j}_+(\Sigma_0,\Sigma_{k'})$, and $N^{s,j}_+(\Sigma_0,D)$.

As is easily verified, for $u \in \mathcal{D}'^{s,j}$, $u \in N^{s,j}(\Sigma_0, \Sigma_{k'})$ tangentially microlocally at (t,x,τ,ξ') if and only if $M_1 \cdots M_{j'} u \in H^s_{ml}(t,x,\tau,\xi')$ (as in Definition 5.3) for all vector fields $M_1, \ldots, M_{j'}$, $j' \le j$, which are simultaneously tangent to Σ_0 and $\Sigma_{k'}$. Distributions in $\mathcal{D}'^{s,j}_+$ which are tangentially microlocally in $N^{s,j}_+(\Sigma_0, \Sigma_1, \ldots, \Sigma_k)$ at (t,x,τ,ξ') may be extended to $\mathcal{D}'^{s,j}$ without losing tangential microlocal regularity.

Lemma 6.15. *Let* $u \in \mathcal{D}'^{s,j}_+$ *and let* $(t_0, 0, x'_0, \tau_0, \xi'_0) \in T^*(\mathbf{R} \times \partial\mathbf{R}^n_+)$. *If* $u \in N^{s,j}_+(\Sigma_0, \Sigma_1, \ldots, \Sigma_k)$ *tangentially microlocally at* $(t_0, 0, x'_0, \tau_0, \xi'_0)$, *then* u *can be extended to* $v \in \mathcal{D}'^{s,j}$ *such that* $v \in N^{s,j}(\Sigma_0, \Sigma_1, \ldots, \Sigma_k)$ *tangentially microlocally at* $(t_0, 0, x'_0, \tau_0, \xi'_0)$.

Proof. We may write

$$u = p(t,x,D_t,D_{x'})u + (1 - p(t,x,D_t,D_{x'}))u,$$

with $p(t,x,D_t,D_{x'})u \in N^{s,j}_+(\Sigma_0, \Sigma_1, \ldots, \Sigma_k)$ and $p(t,x,\tau,\xi') = 1$ on a conic neighborhood of $(t_0, 0, x'_0, \tau_0, \xi'_0)$. Then $p(t,x,D_t,D_{x'})u$ can be extended to an element $w \in N^{s,j}(\Sigma_0, \Sigma_1, \ldots, \Sigma_k)$. On the other hand, as in Lemma 6.9, u may be extended to $u' \in \mathcal{D}'^{s,j}$. If $q(t,x,\tau,\xi')$ has support in a sufficiently small conic neighborhood of $(t_0, 0, x'_0, \tau_0, \xi'_0)$, then

$$q(t,x,D_t,D_{x'})(1 - p(t,x,D_t,D_{x'}))u' \in N^{s,j}(\Sigma_0, \Sigma_1, \ldots, \Sigma_k)$$

by (6.6) and Lemma 6.13. Therefore $v = w + (1 - p(t,x,D_t,D_{x'}))u'$ is an extension of u of the desired form. Q.E.D.

Tangentially microlocally away from the conormal to Δ, it will not be necessary to use fully pseudodifferential operators in order to separate the conormals to the surfaces $\Sigma_1, \ldots, \Sigma_k$. In fact, tangentially microlocally away from the conormal to Δ, as in Figure 6.2, only regularity of type $N^{s,j}_+(\Sigma_0, \Delta)$ or $N^{s,j}(\Sigma_0, \Delta)$ needs to be examined.

Lemma 6.16. *Let* $u \in \mathcal{D}'^{s,j}_+$ *and let* $(t_0, x_0, \tau_0, \xi'_0) \in \mathbf{R}^{n+1} \times \mathbf{R}^n \setminus 0$ *be a point such that* $(t_0, x_0, \tau_0, \xi_1, \xi'_0) \notin N^*(\Sigma_{k'})$, $1 \le k' \le k$, *for all* $\xi_1 \in \mathbf{R}$. *Then* $u \in N^{s,j}_+(\Sigma_0, \Sigma_1, \ldots, \Sigma_k)$ *tangentially microlocally at* $(t_0, x_0, \tau_0, \xi'_0)$ *if and only if* $u \in N^{s,j}_+(\Sigma_0, \Delta)$ *tangentially microlocally at* $(t_0, x_0, \tau_0, \xi'_0)$.

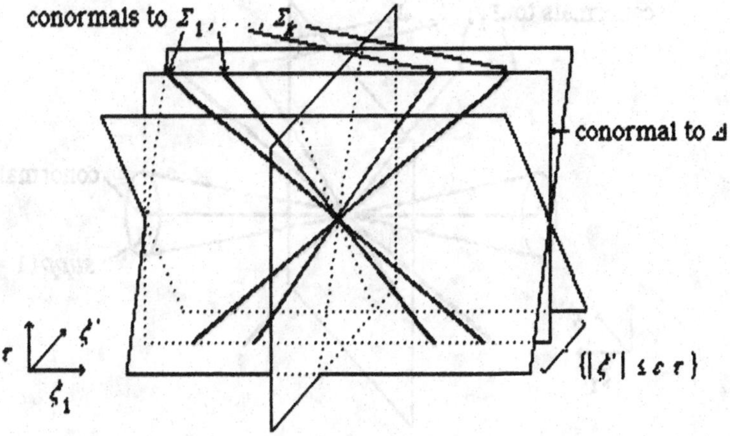

Figure 6.2

Proof. Since $N^{s,j}_+(\Sigma_0, \Delta) \subset N^{s,j}_+(\Sigma_0, \Sigma_{k'})$ for each k', it is a consequence of Definition 6.9 and Lemma 6.13 that $u \in N^{s,j}_+(\Sigma_0, \Delta)$ tangentially microlocally at $(t_0, x_0, \tau_0, \xi'_0)$ if $u \in N^{s,j}_+(\Sigma_0, \Sigma_1, \ldots, \Sigma_k)$ tangentially microlocally at $(t_0, x_0, \tau_0, \xi'_0)$. Conversely, suppose that $u \in N^{s,j}_+(\Sigma_0, \Sigma_1, \ldots, \Sigma_k)$ tangentially microlocally at $(t_0, x_0, \tau_0, \xi'_0)$. If $(t_0, x_0, \tau_0, \xi'_0) \in T^*(\mathbf{R} \times \partial\mathbf{R}^n)$, let v be an extension of u as in Lemma 6.15; otherwise let $v = u$.

Let $p(t, x, D_t, D_{x'})$ be a boundary pseudodifferential operator of order zero, microlocally elliptic at $(t_0, x_0, \tau_0, \xi'_0)$, such that $p(t, x, D_t, D_{x'}) v \in N^{s,j}(\Sigma_0, \Sigma_1, \ldots, \Sigma_k)$. If $q(t, x, \tau, \xi) \in S^0_{1,0}(\mathbf{R}^{n+1})$ is supported in $\{(\tau, \xi): |(\tau, \xi')| \ge \varepsilon|\xi_1|\}$ and is identically one for $|(\tau, \xi')| \ge 2\varepsilon|\xi_1|$ and $|(\tau, \xi)| \ge 1$, then for ε sufficiently small, the support of $(1 - q)$ is near the conormal to Σ_0, as in Figure 6.3.

It follows from the proof of Lemma 6.6 that

$$(1 - q(t, x, D_t, D_{x'})) p(t, x, D_t, D_{x'}) v \in N^{s,j}(\Sigma_0, \Delta).$$

On the other hand, $q(t, x, D_t, D_{x'}) p(t, x, D_t, D_{x'})$ is an operator with symbol in $S^0_{1,0}(\mathbf{R}^{n+1})$, and since the support of its symbol is located away from the union of the conormals to $\Sigma_0, \Sigma_1, \ldots, \Sigma_k$, it follows from Lemma 6.6 that

$$q(t, x, D_t, D_{x'}) p(t, x, D_t, D_{x'}) v \in N^{s,j}(\Sigma_0, \Delta).$$

Therefore, $p(t, x, D_t, D_{x'}) v \in N^{s,j}(\Sigma_0, \Delta)$, $p(t, x, D_t, D_{x'}) u \in N^{s,j}_+(\Sigma_0, \Delta)$, and $u \in N^{s,j}_+(\Sigma_0, \Delta)$ tangentially microlocally at $(t_0, x_0, \tau_0, \xi'_0)$. Q.E.D.

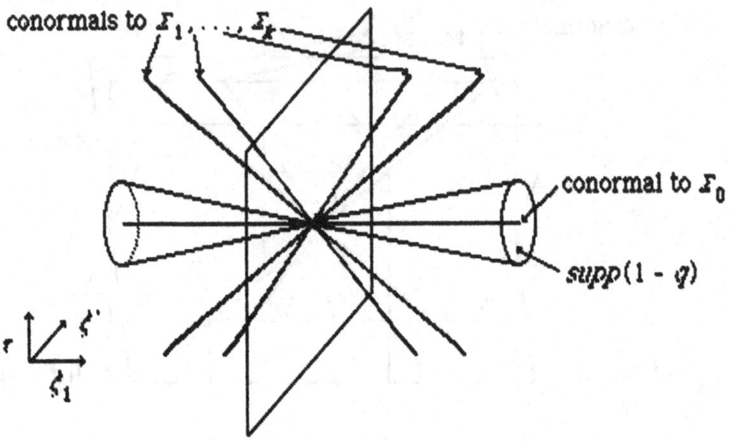

Figure 6.3

The results on the propagation of conormal regularity in the presence of more than two smooth characteristic hypersurfaces in the reflected family for a semilinear boundary problem may now be stated.

Theorem 6.17. *Let $p(t,x,D)$ be a strictly hyperbolic operator of order m and let $B = (B_j)_{j=0,\ldots,\mu}$ be a collection of boundary operators satisfying* (6.3). *Let Σ_1 be a smooth characteristic hypersurface for p_m which intersects $\Sigma_0 = \mathbf{R}\times\partial\mathbf{R}^n_+$ transversally in the smooth submanifold Δ in $\{t \geq 0\}$, such that* (6.4) *holds. Let $\{\Sigma_1, \ldots, \Sigma_k\}$ be the corresponding reflected family. Suppose that $u \in H^s_{loc}(\mathbf{R}\times\partial\mathbf{R}^n_+)$, $s > (n+1)/2 + m - 2$, f is smooth, and u satisfies $p(t,x,D)u = f(t,x,u,\ldots,D^{m-2}u)$ on $\mathbf{R}\times\partial\mathbf{R}^n_+$, $Bu|_{x_1=0} = 0$. If for some $\varepsilon > 0$, $u \in N^{s,j}_+(\Sigma_1\cap\{t<-\varepsilon\},\ldots,\Sigma_k\cap\{t<-\varepsilon\})$, then $u \in N^{s,j}_+(\Sigma_0,\Sigma_1,\ldots,\Sigma_k)$, and $(\partial x_1)^i u|_{x_1=0} \in N^{s-i}(\Delta)$, $0 \leq i \leq m-1$.*

The proof involves separate arguments on a tangential microlocal neighborhood of $N^*(\Sigma_1)\cup\ldots\cup N^*(\Sigma_k)$ and away from that set, as in Figure 6.4. Near the conormals, the equation may be reduced to a first order pseudodifferential system with forward and backward elliptic and hyperbolic parts, as in Taylor [70]. Let v be a component of the corresponding system satisfying the hyperbolic equation

$$(D_1 - \lambda(t,x,D_t,D_{x'}))v = f \text{ on } \mathbf{R}\times\mathbf{R}^n_+.$$

If $f \in \mathcal{D}'^{s,j}_+$ and, inductively, $f \in N^{s,j}_+(\Sigma_0,\Sigma_1,\ldots,\Sigma_k)$ microlocally at

$(t_0, 0, x'_0, \tau_0, \xi'_0)$, then by Lemma 6.15, f may be extended across the boundary to an element of $\mathcal{D}'^{s,j}$ such that $f \in N^{s,j}_+(\Sigma_0, \Sigma_1, \ldots, \Sigma_k)$ microlocally at $(t_0, 0, x'_0, \tau_0, \xi'_0)$. If w is the unique solution on $\mathbf{R} \times \mathbf{R}^n$ to

$$(D_1 - \lambda(t, x, D_t, D_{x'}))w = f \text{ on } \mathbf{R} \times \mathbf{R}^n_+,\ w(t, 0, x') = v(t, 0, x'),$$

then w is an extension of v, and $w \in \mathcal{D}'^{s+1,j}$ by Lemma 5.2. Fully pseudodifferential operators may be applied in the microlocal analysis of w. Conormal regularity for these components is a consequence of the inductive hypotheses for f and the trace of w on $\mathbf{R} \times \partial \mathbf{R}^n_+$, and the following analogues of Lemmas 5.6 and 5.7.

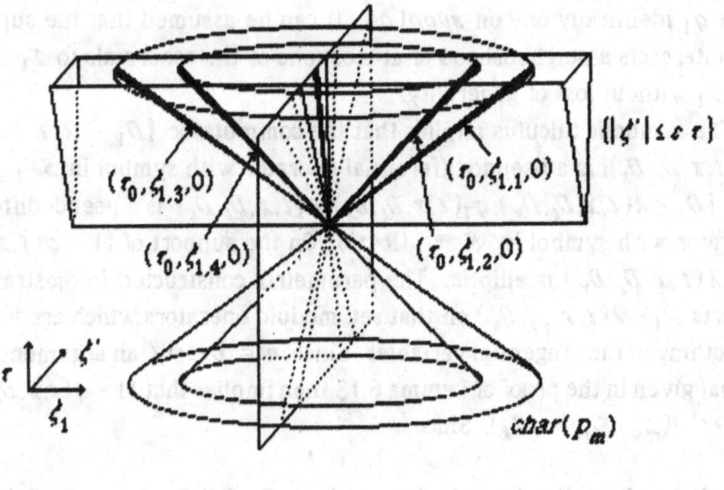

Figure 6.4

Lemma 6.18. *Let $\lambda(t, x, D_t, D_{x'})$ be a classical properly supported pseudodifferential operator of order one with real principal symbol. Let Γ be a null bicharacteristic in $\mathbf{R} \times \mathbf{R}^n$ for $\xi_1 - \lambda(t, x_1, x', \tau, \xi')$ which passes through $\gamma_0 = (t_0, 0, x'_0, \tau_0, \xi_1, \xi'_0)$ and $\gamma_1 = (s_0, y_1, y'_0, \sigma_0, \eta_1, \eta'_0)$. Assume that $w \in \mathcal{D}'^{s+1,j}$, $f \in \mathcal{D}'^{s,j}$, $f \in N^{s,j}(\Sigma_0, \Sigma_1, \ldots, \Sigma_k)$ tangentially microlocally at all points (t, x_1, x', τ, x') for which $(t, x_1, x', \tau, \xi_1, \xi') \in \Gamma$, and*

$$(D_1 - \lambda(t, x, D_t, D_{x'}))w = f.$$

If $w \in N^{s,j}(\Sigma_0, \Sigma_1, \ldots, \Sigma_k)$ tangentially microlocally at $(s_0, y_1, y'_0, \sigma_0, \eta'_0)$, then $w \in N^{s,j}(\Sigma_0, \Sigma_1, \ldots, \Sigma_k)$ tangentially microlocally at $(t_0, 0, x'_0, \tau_0, \xi'_0)$.

and $w(t,0,x') \in N^{s,j}(\Delta)$ *microlocally at* $(t_0,0,x'_0,\tau_0,\xi'_0)$.

Lemma 6.19. *Let* $\lambda(t,x,D_t,D_{x'})$, Γ, γ_0, *and* γ_1 *be as in Lemma* 6.18. *Let* $w \in \mathcal{D}'^{s+1,j}$ *and suppose that* $(D_1 - \lambda(t,x,D_t,D_{x'}))\,w \in N^{s,j}(\Sigma_0,\Sigma_1,\ldots,\Sigma_k)$ *tangentially microlocally at all points* (t,x_1,x',τ,ξ') *with* $(t,x_1,x',\tau,\xi_1,\xi') \in \Gamma$. *If* $w(t,0,x') \in N^{s,j}(\Delta)$ *microlocally at* $(t_0,0,x'_0,\tau_0,\xi'_0)$, *then* $w \in N^{s,j}(\Sigma_0,\Sigma_1,\ldots,\Sigma_k)$ *tangentially microlocally at* $(t_0,0,x'_0,\tau_0,\xi'_0)$ *and at* $(s_0,y_1,y'_0,\sigma_0,\eta'_0)$.

Proof. Let $q(t,x,\tau,\xi) \in S^0_{1,0}(\mathbf{R}^{n+1})$ have symbol which is supported in a small conic neighborhood of $\{\xi_1 = \lambda(t,x_1,x',\tau,\xi')\}$, with q identically one on a smaller neighborhood of that set. Let $q_1(t,x,\tau,\xi)$ have similar properties, with q_1 identically one on $supp(q)$. It can be assumed that the support of q_1 intersects a neighborhood of at most one of the conormals to $\Sigma_1,\ldots,\Sigma_k$, say Σ_1 without loss of generality.

The symbolic calculus implies that the commutator $[D_1 - \lambda(t,x,D_t,D_{x'}),\ q_1(t,x,D_t,D_x)]$ is a pseudodifferential operator with symbol in $S^0_{1,0}(\mathbf{R}^{n+1})$, and $[D_1 - \lambda(t,x,D_t,D_{x'}), q_1(t,x,D_t,D_x)]\,q(t,x,D_t,D_x)$ is a pseudodifferential operator with symbol in $S^{-\infty}_{1,0}(\mathbf{R}^{n+1})$. On the support of $(1 - q(t,x,t,x))$, $D_1-\lambda(t,x,D_t,D_{x'})$ is elliptic. The parametrix constructed in Sjöstrand [68] inverts $D_1 - \lambda(t,x,D_t,D_{x'})$ on that set, modulo operators which are infinitely smoothing in the tangential variables. Since $w \in \mathcal{D}'^{s+1,j}$, an argument similar to that given in the proof of Lemma 6.13 then implies that $(1 - q(t,x,D_t,D_x))\,w \in N^{s+1,j}(\Sigma_0,\Sigma_1,\ldots,\Sigma_k)$. Since

$$
\begin{aligned}
(D_1-\lambda(t,x,D_t,D_{x'}))\,q_1(t,x,D_t,D_x) &= q_1(t,x,D_t,D_x)(D_1-\lambda(t,x,D_t,D_{x'}))\\
&\quad + [D_1-\lambda(t,x,D_t,D_{x'}),q_1(t,x,D_t,D_x)](1 - q(t,x,D_t,D_x))\\
&\quad + [D_1-\lambda(t,x,D_t,D_{x'}),q_1(t,x,D_t,D_x)]\,q(t,x,D_t,D_x),
\end{aligned}
$$

it follows that $(D_1 - \lambda(t,x,D_t,D_{x'}))\,q_1(t,x,D_t,D_x)\,w \in N^{s,j}(\Sigma_0,\Sigma_1)$. The proof of Theorem 3.3 may be easily adapted to yield the desired local conclusions about $q_1(t,x,D_t,D_x)\,w$ and its trace on $\mathbf{R}\times\partial\mathbf{R}^n_+$, and therefore the microlocal statements about w and $w(t,0,x')$ hold. See Beals-Métivier [12] for additional details. Q.E.D.

The components of the pseudodifferential system corresponding to solutions of forward and backward elliptic equations do not extend across the boundary as solutions to an elliptic problem, and therefore fully pseudodifferential operators may not be applied to them. However, the elliptic operators may be inverted modulo tangentially smoothing operators, so by (6.6) the analysis of the conormal regularity of these pieces is reduced to a proof of regularity of the type $N^{s,j}_{+}(\Sigma_0,\Delta)$. Since vector fields alone are used to define $N^{s,j}_{+}(\Sigma_0,\Delta)$, induction on the interior estimate and the usual microlocal elliptic regularity theorem are enough to treat the forward elliptic part. Interior regularity in combination with the inductive hypothesis on the boundary and Lemma 6.11 allows the backward elliptic part to be handled similarly. The precise results which are used may be stated as follows. See Beals-Métivier [12] for the proofs.

Lemma 6.20. *Let* $e(t,x_1,x',\tau,\xi')\in S^1_{1,0}(\mathbf{R}^{n+1})$ *be a square matrix of symbols with all of the eigenvalues of* $e(t_0,0,x'_0,\tau_0,\xi'_0)$ *located in the upper half plane* $Im(\lambda)>0$. *Suppose that* $w\in \mathcal{D}'^{s+1,j}_{+}$ *for some* $s>0$, $f\in \mathcal{D}'^{s,j}_{+}$, $f\in N^{s,j}_{+}(\Sigma_0,\Sigma_1,\ldots,\Sigma_k)$ *tangentially microlocally at* $(t_0,0,x'_0,\tau_0,\xi'_0)$, *and* $(D_1-e(t,x,D_t,D_{x'}))w = f$. *If* $w(t,0,x')\in N^{s,j}(\Delta)$ *microlocally at* $(t_0,0,x'_0,\tau_0,\xi'_0)$, *then* $w\in N^{s+1/2,j}_{+}(\Sigma_0,\Sigma_1,\ldots,\Sigma_k)$ *tangentially microlocally at* $(t_0,0,x'_0,\tau_0,\xi'_0)$.

Lemma 6.21. *Let* $e(t,x_1,x',\tau,\xi')\in S^1_{1,0}(\mathbf{R}^{n+1})$ *be a square matrix of symbols with all of the eigenvalues of* $e(t_0,0,x'_0,\tau_0,\xi'_0)$ *located in the lower half plane* $Im(\lambda)<0$. *Suppose that* $w\in \mathcal{D}'^{s+1,j}_{+}$ *for some* $s>0$, $f\in \mathcal{D}'^{s,j}_{+}$, $f\in N^{s,j}_{+}(\Sigma_0,\Sigma_1,\ldots,\Sigma_k)$ *tangentially microlocally at* $(t_0,0,x'_0,\tau_0,\xi'_0)$, *and* $(D_1-e(t,x,D_t,D_{x'}))w = f$. *Then* $w\in N^{s+1,j}_{+}(\Sigma_0,\Sigma_1,\ldots,\Sigma_k)$ *tangentially microlocally at* (t_0,x_0,τ_0,ξ'_0). *If* $x_0=(0,x'_0)$, *then* $w(t,0,x')\in N^{s+1/2,j}(\Delta)$ *microlocally at* $(t_0,0,x'_0,\tau_0,\xi'_0)$.

Tangentially microlocally away from $N^*(\Sigma_1)\cup\ldots\cup N^*(\Sigma_k)$, by Lemma 6.16 we have $N^{s,j}_{+}(\Sigma_0,\Sigma_1,\ldots,\Sigma_k) = N^{s,j}_{+}(\Sigma_0,\Delta)$, and therefore fully pseudodifferential operators are again not needed on that set to establish the desired regularity in Theorem 6.17. In the local coordinates for which $\Delta = \{t = x_1 = 0\}$, the vector fields simultaneously tangent to Σ_0 and Δ are generated by $\{t\partial_t, x_1\partial_t, x_1\partial x_1, \partial x_2,\ldots,\partial x_n\}$. The (τ,ξ) projection of the set in question is $\{(\tau,\xi_1,\xi'): |\xi'|\geq \varepsilon|\tau|\}$, and on that set $\{x_1\partial x_1,\partial x_2,\ldots,\partial x_n\}$ is therefore a collection of generators. Since locally after division by a nonzero function, $p_m(t,x,D) = (\partial x_1)^n + r(t,x,D)$, with r of lower order in ∂x_1, it easily follows that these vector fields satisfy appropriate commutation properties with p_m. Thus conormal regularity for the solution u is established by induction on j. See Beals-Métivier [12] for the remaining details of this argument.

The propagation of conormal regularity with respect to a reflected pair of nongrazing characteristic hypersurfaces for the solution of the general semilinear strictly hyperbolic equation $p_m(t,x,D)u = f(t,x,u,\ldots,D^{m-1}u)$ on $\mathbf{R}\times\mathbf{R}^n_+$, under the hypothesis $s > (n+1)/2 + m - 1$, has been established by Wang [72].

The more difficult case of the reflection of conormal singularities when grazing directions are present remains open, even for a simple second order problem of the form

$$\Box u = f(t,x,u) \text{ on } \mathbf{R}\times\Omega,\ u|_{\mathbf{R}\times\partial\Omega} = 0,$$

where Ω is the exterior of the unit ball in $\mathbf{R}^n$ and u is assumed to be conormal with respect to a single characteristic hyperplane in the past. It is expected that the grazing points will be sources of nonlinear conormal singularities, as illustrated in Figure 6.5.

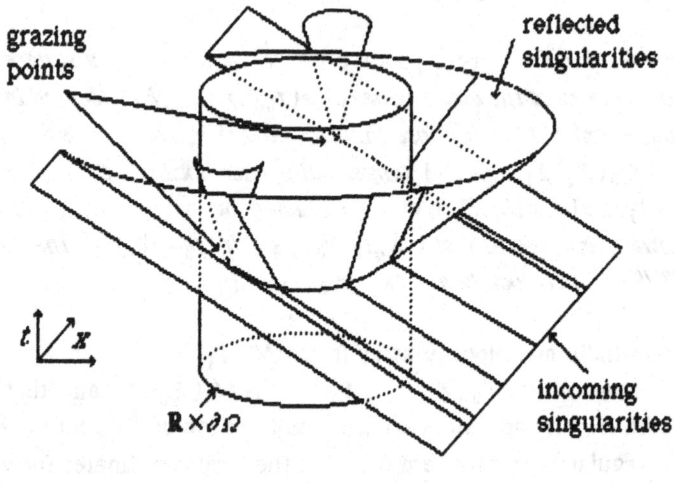

Figure 6.5

When a pair of incoming transversal characteristic hypersurfaces for a second order equation intersect at the boundary, the family consisting of the incoming and reflected hypersurfaces and the boundary has five elements. Thus it is expected from the interior example discussed in Chapter III that solutions to nonlinear problems with conormal singularities in the past on the incoming pair will in general have conormal singularities on the outgoing pair and on the surface of the light cone over the interaction point, as in Figure 6.6 a). This cone will in general have a direction grazing the boundary, so the analysis of such an interaction will be complicated. Chen [28] has considered the case of the d'Alembertian □ and the flat boundary, as in Figure 6.6 b).

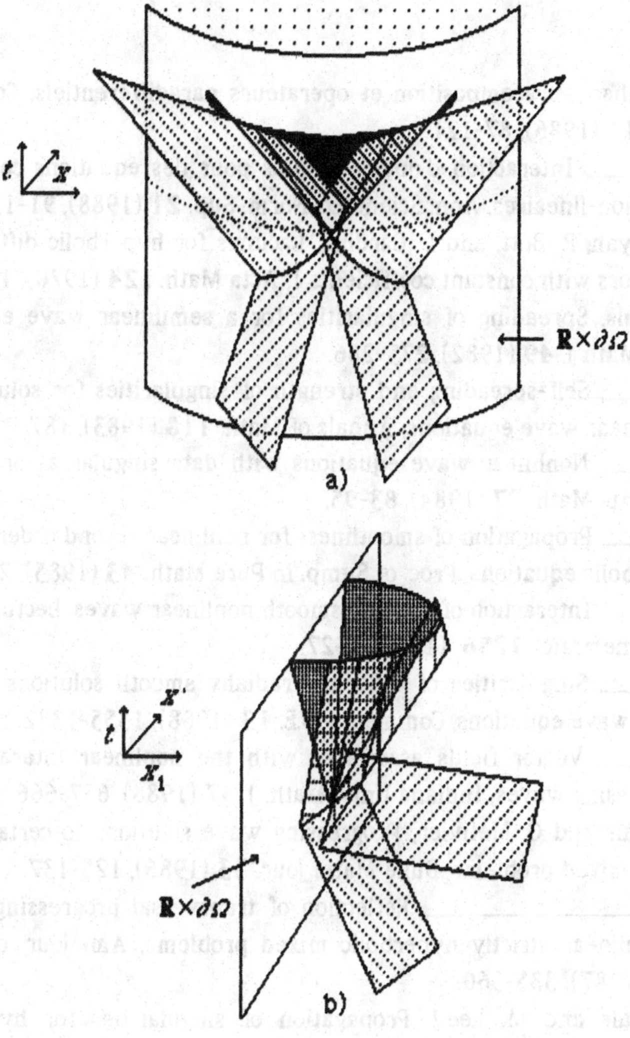

Figure 6.6

Bibliography

[1] S. Alinhac, Paracomposition et operateurs paradifferentiels, Comm. in P.D.E. **11** (1986), 87-121.

[2] ________, Interaction d'ondes simples pour des équations complètement non-lineaires, Ann. Scient. Ec. Norm. Sup. **21** (1988), 91-132.

[3] M. Atiyah, R. Bott, and L. Gårding, Lacunas for hyperbolic differential operators with constant coefficients I, Acta Math. **124** (1970), 109-189.

[4] M. Beals, Spreading of singularities for a semilinear wave equation, Duke Math J. **49** (1982), 275-286.

[5] ________, Self-spreading and strength of singularities for solutions to semilinear wave equations, Annals of Math. **118** (1983), 187-214.

[6] ________, Nonlinear wave equations with data singular at one point, Contemp. Math. **27** (1984), 83-95.

[7] ________, Propagation of smoothness for nonlinear second order strictly hyperbolic equations, Proc. of Symp. in Pure Math. **43** (1985), 21-44.

[8] ________, Interaction of radially smooth nonlinear waves, Lecture Notes in Mathematics **1256** (1987), 1-27.

[9] ________, Singularities of conormal radially smooth solutions to nonlinear wave equations, Comm. in P.D.E. **13** (1988), 1355-1382.

[10] ________, Vector fields associated with the nonlinear interaction of progressing waves, Indiana Univ. Math. J. **37** (1988), 637-666.

[11] M. Beals and G. Métivier, Progressing wave solutions to certain nonlinear mixed problems, Duke Math. Jour. **53** (1986), 125-137.

[12] ____________________, Reflection of transversal progressing waves in nonlinear strictly hyperbolic mixed problems, Am. Jour. of Math. **109** (1987), 335-360.

[13] M. Beals and M. Reed, Propagation of singularities for hyperbolic pseudodifferential operators and applications to nonlinear problems, Comm. Pure Appl. Math **35** (1982), 169-184.

[14] ______________, Microlocal regularity theorems for nonsmooth pseudodifferential operators and applications to nonlinear problems, Trans. Am. Math. Soc. **285** (1984), 159-184.

[15] J.-M. Bony, Calcul symbolique et propagation des singularités pour les équations aux dérivées partielles non-linéaires, Ann. Scien. École Norm. Sup. **14** (1981), 209-246.

[16] ______, Interaction des singularités pour les équations aux d'rivées partielles non-linéaires, Sem. Goulaouic-Meyer-Schwartz, exp. no. 22 (1979-80).

[17] ______, Interaction des singularités pour les équations aux dérivées partielles non-linéaires, Sem. Goulaouic-Meyer-Schwartz, exp. no. 2 (1981-82).

[18] ______, Interaction des singularités pour les équations de Klein-Gordon nonlinéaires, Sem. Goulaouic-Meyer-Schwartz, exp. no. 10 (1983-84).

[19] ______, Second microlocalization and propagation of singularities for semilinear hyperbolic equations, Taniguchi Symp., Katata (1984), 11-49.

[20] J.-M. Bony and N. Lerner, Quantification asymptotique et microlocalisations d'ordre superieur, Sem. Equations aux Derivees Partielles, exp. no. 2 and 3 (1986-1987).

[21] J. Chazarain and A. Piriou, *Introduction to the Theory of Linear Partial Differential Equations*, North-Holland, Amsterdam, (1982).

[22] J.-Y. Chemin, Interaction de trois ondes dans les équations semilinéaires strictement hyperboliques d'ordre 2, These Universite de Paris Centre D'Orsay (1987).

[23] ______, Calcul paradifferentiel précisé et applications à des équations aux dérivées partielles non semi linéaires, Duke Math. J. **56** (1988), 431-469.

[24] ______, Évolution d'une singularité ponctuelle dans des équations strictement hyperboliques non linéaires, preprint.

[25] S. Chen, Pseudodifferential operators with finitely smooth symbols and their applications to quasilinear equations, Nonlin. Analysis, Theory, Methods., and Appl. **6** (1982), 1193-1206.

[26] ______, Regularity estimate of solution to semilinear wave equation in higher space dimension, Acta Scientia Sinica **27** (1984), 924-935.

[27] ______, The reflection and interaction of the singularities of solutions to semilinear wave equation in higher space dimension, Nonlin. Analysis,

Theory, Methods, and Appl. **8** (1984), 1167-1179.

[28] ______, Reflection and interaction of progressing waves for semilinear wave equations, Research report, Nankai Institute of Math. (1986).

[29] R. R. Coifman and Y. Meyer, Au-delà des opérateurs pseudo-différentiels, Astérisque, Soc. Math. France **57**, 1978.

[30] F. David and M. Williams, Singularities of solutions to semilinear boundary value problems, Am. Jour. of Math. **109** (1987), 1087-1109.

[31] J. M. Delort, Deuxieme microlocalisation simultanee et front d'onde de produits, preprint.

[32] G. Folland, *Lectures on Partial Differential Equations*, Tata Institute Lectures, Springer-Verlag, Berlin, (1983).

[33] P. Gerard and J. Rauch, Propagation de la regularite locale de solutions d'équations hyperboliques non linéaires, preprint.

[34] L. Holt, Rutgers University Thesis, in preparation.

[35] L. Hörmander, Fourier integral operators I, Acta Math. **127** (1971), 79-183.

[36] ______, Spectral analysis of singularities, Annals of Math. Studies **91** (1979), 3-49.

[37] ______, *The Analysis of Linear Partial Differential Operators I, II, III, IV*, Springer-Verlag, Berlin, (1983,1985).

[38] S. Kang, Rutgers University Thesis, in preparation.

[39] Y. Laurent, Théorie de la deuxième *microlocalisation* dans le domaine complexe, Progress in Math. **53**, Birkhäuser, (1985).

[40] P. Lax, Hyperbolic systems of conservation laws and the mathematical theory of shock waves, C.B.M.S. Reg. Conf. Ser. in Math. **11**, Soc. Ind. Appl. Math., (1973).

[41] G. Lebeau, Problème de Cauchy semi-linéaire en 3 dimensions d'espace. Un résultat de finitude, Jour. Func. Anal. **78** (1988), 185-196.

[42] ______, Equations des ondes semi-linéaires II. Controle des singularités et caustiques non linéaires, preprint.

[43] E. Leichtnam, Régularité microlocale pour des problèmes de Dirichlet nonlinéaires noncaractéristiques d'ordre deux à bord peu regulier, Bull. Soc. Math. France **115** (1987), 457-489.

[44] L. Liu, Optimal propagation of singularities for semilinear hyperbolic differential equations, Fudan University Thesis (1984).

[45] R. Melrose, Semilinear waves with cusp singularities, preprint.

[46] R. Melrose and N. Ritter, Interaction of nonlinear progressing waves, Annals of Math. **121** (1985), 187-213.

[47] ________, Interaction of progressing waves for semilinear wave equations II, Arkiv for Mat. **25** (1987), 91-114.

[48] R. Melrose and J. Sjöstrand, Singularities of boundary value problems I, Comm. Pure Appl. Math. **31** (1978), 593-617.

[49] T. Messer, Propagation of singularities of hyperbolic systems, Indiana Univ. Math. J. **36** (1987), 45-77.

[50] G. Métivier, The Cauchy problem for semilinear hyperbolic systems with discontinuous data, Duke Math. J. **53** (1986), 983-1011.

[51] G. Métivier and J. Rauch, Existence and interaction of piecewise smooth waves for first order semilinear systems, preprint.

[52] Y. Meyer, Remarque sur un théorème de J.-M. Bony, Suppl. Rend. Circ. Mat. Palermo **1** (1981), 1-20.

[53] B. Nadir and A. Piriou, Ondes semi-linéaires conormales par rapport á deux hypersurfaces transverses, preprint.

[54] L. Nirenberg, Lectures on linear partial differential equations, C.B.M.S. Reg. Conf. Ser. in Math. **17**, Am. Math. Soc., (1973).

[55] A. Piriou, Calcul symbolique non linéaire pour une onde conormale simple, preprint.

[56] Q. Qiu, Para-Fourier integral operators, preprint.

[57] J. Rauch, Singularities of solutions to semilinear wave equations, J. Math. Pures et Appl. **58** (1979), 299-308.

[58] J. Rauch and M. Reed, *Propagation of singularities for semilinear* hyperbolic equations in one space variable, Ann. of Math. **111** (1980), 531-552.

[59] ________, Nonlinear microlocal analysis of semilinear hyperbolic systems in one space dimension, Duke Math. J. **49** (1982), 379-475.

[60] ________, Singularities produced by the nonlinear interaction of three progressing waves: examples, Comm. in P.D.E. **7** (1982), 1117-1133.

[61] ________, Discontinuous progressing waves for semilinear systems, Comm. in P.D.E. **10** (1985), 1033-1075.

[62] ________, Striated solutions of semilinear two-speed wave equations, Indiana Univ. Math. J. **34** (1985), 337-353.

[63] ________, Classical conormal solutions of semilinear systems, Comm. in P.D.E. **13** (1988), 1297-1335.

[64] N. Ritter, Progressing wave solutions to nonlinear hyperbolic Cauchy problems, M. I. T. Thesis (1984).

[65] A. Sá Barreto, Interactions of conormal waves for fully semilinear wave equations, Am. Jour. of Math., to appear.

[66] A. Sá Barreto and R. Melrose, Examples of non-discreteness for the interaction geometry of semilinear progressing waves in two space dimensions, preprint.

[67] M. Sablé-Tougeron, Régularité microlocale pour des problemes aux limites non linéaires, Ann. Inst. Fourier **36** (1986), 39-82.

[68] J. Sjöstrand, Operators of principal type with interior boundary conditions, Acta Math. **130** (1973), 1-51.

[69] E. Stein, *Singular Integrals and Differentiability Properties of Functions*, Princeton University Press, Princeton, (1970).

[70] M. Taylor, *Pseudodifferential Operators*, Princeton University Press, Princeton, (1981).

[71] F. Treves, *Basic Linear Partial Differential Equations*, Academic Press, New York, (1975).

[72] Y. Wang, Singularities produced by the reflection and interaction of two progressing waves, preprint.

[73] M. Williams, Spreading of singularities at the boundary in semilinear hyperbolic mixed problems I: microlocal $H^{s,s'}$ regularity, Duke Math. J. **56** (1988), 17-40.

[74] ________, Spreading of singularities at the boundary in semilinear hyperbolic mixed problems II: crossing and self-spreading, Trans. Am. Math. Soc., to appear.

[75] C. Xu, Propagation au bord des singularités pour des problemes de Dirichlet non linéaires d'ordre deux, preprint.

Index

Index of Notation

$C(\mathbf{R}; H^s_{loc}(\mathbf{R}^n))$ 64
$C^\infty_{com}(\mathbf{R}^{n+1})$ 4

$\mathcal{D}^{s,j}$ 126
$\mathcal{D}^{s,j}_+$ 126
$\mathcal{D}'^{s,j}$ 126
$\mathcal{D}'^{s,j}_+$ 126
$\Delta_k V$ 21
δ_0 4

$H^{n/2-}$ 82
H_{p_2} 109
$H(r)$ 4
$H^s(\mathbf{R}^n)$ 10
$H^s(\mathbf{R}^n_+)$ 100
$H^s_{loc}(\mathbf{R}^n)$ 10
$H^s_{ml}(x_0, \xi_0)$ 12
$H^{s,r,g}$ 36
$H^{s,r,g}(\Gamma)$ 40
$H^{s,s'}(\mathbf{R}^n)$ 100
$H^{s,s'}(\mathbf{R}^n_+)$ 100
$H^{s,s'}_{ml}(0, x'_0, \xi'_0)$ 111
$H^{s-;k_1,k_2,k_3}$ 77

$L^p(\mathbf{R}^n)$ 61

$N^{s,j}(\Sigma_0, \Delta)$ 124
$N^{s,j}_+(\Sigma_0, \Delta)$ 124
$N^{s,j}_+(\Sigma_0, \Sigma_k)$ 124
$N^{s,j}_+(\Sigma_0, \Sigma_1, \ldots, \Sigma_k)$ 125
$N^{s,k}(\Sigma)$ 52, 69
$N^{s,k}(\Sigma_1, \Sigma_2)$ 56
$N^{s,k}(\Sigma_1, \ldots, \Sigma_m)$ 59, 123
$N^{s,k}(\Sigma^-, \Sigma^+, \mathbf{R} \times \partial\mathbf{R}^n_+)$ 119
$N^{3/2-,k}(\{x=0\}, \{y=0\}, \{z=0\}, \{\alpha=0\})$ 82
$N^*(\Sigma)$ 53

$p(x, f, v, D)$ 22
$S_k V$ 22
$S^m_{1,0}$ 13
$S^m_{1,1}$ 22

$T^*(\mathbf{R}^n)$ 8

$WF(u)$ 8

χ_r 8

$\Box$ 4, 75
$\langle \xi \rangle$ 6

Progress in Nonlinear Differential Equations and Their Applications

Editor
Haim Brezis
Department of Mathematics
Rutgers University
New Brunswick, NJ 08903
U.S.A.
and
Département de Mathématiques
Université P. et M. Curie
4, Place Jussieu
75252 Paris Cedex 05
France

Progress in Nonlinear Differential Equations and Their Applications is a book series that lies at the interface of pure and applied mathematics. Many differential equations are motivated by problems arising in diversified fields such as Mechanics, Physics, Differential Geometry, Engineering, Control Theory, Biology, and Economics. This series is open to both the theoretical and applied aspects, hopefully stimulating a fruitful interaction between the two sides. It will publish monographs, polished notes arising from lectures and seminars, graduate level texts, and proceedings of focused and refereed conferences.

We encourage preparation of manuscripts in some such form as LaTex or AMS TEX for delivery in camera ready copy, which leads to rapid publication, or in electronic form for interfacing with laser printers or typesetters.

Proposals should be sent directly to the editor or to: Birkhäuser Boston, 675 Massachusetts Avenue, Suite 601, Cambridge, MA 02139.